GW01418889

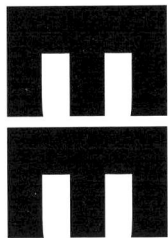

ELECTRICIAN'S GUIDE TO FIRE DETECTION AND ALARM SYSTEMS

Published by The Institution of Engineering and Technology, London, United Kingdom

The Institution of Engineering and Technology is registered as a Charity in England & Wales (no. 211014) and Scotland (no. SC038698).

The Institution of Engineering and Technology is the new institution formed by the joining together of the IEE (The Institution of Electrical Engineers) and the IIE (The Institution of Incorporated Engineers). The new Institution is the inheritor of the IEE brand and all its products and services, such as this one, which we hope you will find useful. The IEE is a registered trademark of the Institution of Engineering and Technology.

© 2010 The Institution of Engineering and Technology

First published 2010

Copies of this publication may be obtained from:
The Institution of Engineering and Technology
PO Box 96
Stevenage
SG1 2SD, UK
Tel: +44 (0)1438 767328
Email: sales@theiet.org
www.theiet.org/publishing/books/wir-reg

ISBN 978-1-84919-130-2

Typeset in the UK by Carnegie Book Production, Lancaster
Printed in the UK by Polestar Wheatons, Exeter

Contents

Cooperating organisations

The IET acknowledges the contribution made by the following organisations in the preparation of this Guide.

BEAMA Installations Ltd
Eur Ing M.H. Mullins BA CEng FIEE FIIE
P. Sayer IEng MIIE GCGI

British Standards Institution
D. Stead

Department for Communities and Local Government
A. Burd
K. Bromley

Electrical Contractors' Association
G. Digilio IEng FIET ACIBSE MSLL

Electrical Contractors' Association of Scotland (t/a SELECT)
D. Millar IEng MIIE MILE

Electrical Safety Council
J. Bradley BSc CEng MIET FCIBSE

Fire and Security Association
M. Turner

Health and Safety Executive
K. Morton CEng FIET

Institution of Engineering and Technology
J.F. Elliott BSc IEng MIET

Scottish Building Standards Agency

Reviewer
C.S. Todd MSc FIFireE FBEng MIRM MSFPE CPhys FInstP CEng FIET

Author
P.R.L. Cook CEng FIET

The author would like to record special thanks to Eur Ing L. Markwell MSc BSc CEng MIET MCIBSE LCGI of the ECA for his particular assistance in preparing the publication.

Acknowledgements

References to British Standards are made with the kind permission of BSI. Complete copies can be obtained by post from:

BSI Customer Services
389 Chiswick High Road
London W4 4AL

Tel: +44 (0)20 8996 9001
Fax: +44 (0)20 8996 7001
Email: cservices@bsi-global.com
Web: www.bsi-global.com

Department for Communities and Local Government building regulations publications including:

▶ Building Regulations and Fire Safety: Procedural Guidance
▶ Approved Document B: Fire safety

Preface

The *Electrician's Guide to Fire Detection and Alarm Systems* is one of a number of publications prepared by the IET to provide guidance on electrical installations in buildings. This publication is concerned with fire detection and fire alarm systems and must be read in conjunction with the legislation, Approved Document B and the British Standards, in particular BS 5839-1 and 5839-6.

Designers and installers should always consult these documents to satisfy themselves of compliance.

It is expected that persons carrying out work in accordance with this Guide will be competent to do so, competence being a statutory requirement of the Electricity at Work Regulations.

Scope

This guide has been prepared for electrical installation designers and contractors involved in the design, installation and/or maintenance of fire detection and alarm systems in England and Wales. Guidance on the requirements for Scotland is given in Chapter 8, prepared with the assistance of the Scottish Building Standards.

Fire safety legislation

1.1 Introduction

A major change was made to fire safety legislation in England and Wales in 2006 by the Regulatory Reform (Fire Safety) Order 2005 (the 'Fire Safety Order'). The Fire Safety Order is the principal legislation for general fire safety and reforms the law relating to fire safety in non-domestic premises used or operated by employers, the self-employed and the voluntary sector. (Exceptions would be offshore installations, means of transport, mines and boreholes. There are special considerations for these areas and separate regimes for safety, which continue.) Equivalent legislative changes were introduced in Scotland by means of the Fire (Scotland) Act 2005 and the Fire Safety (Scotland) Regulations 2006. Similar changes will be introduced in Northern Ireland in 2009.

1.1.1 Legislation repealed, revoked or amended

Legislation repealed, revoked or amended includes:

- the Fire Precautions Act 1971 (whole act repealed and ceases to have effect)
- any regulations made under the Health and Safety at Work etc. Act 1974 concerning general fire precautions in relation to premises to which the Fire Safety Order applies (amended). That is in so far as the Act etc. imposes requirements or prohibitions concerning general fire precautions
- the Fire Safety and Safety of Places of Sport Act 1987 and the Environment and Safety Information Act 1988 (amended to make reference to the Fire Safety Order)
- the Fire Precautions and Workplace Regulations 1997 (revoked)
- the Management of Health and Safety at Work Regulations (which no longer control general fire precautions).

Much local authority legislation is amended to remove general fire safety provisions included in the Fire Safety Order, such as:

- Safety of Sports Grounds Act 1975
- Greater London Council (General Powers) Act 1975
- County of South Glamorgan Act 1976
- Local Government, Planning and Land Act 1980
- West Midlands County Council Act 1980
- Cheshire County Council Act 1980
- Local Government (Miscellaneous) Provisions Act 1982
- Humberside Act 1982

The Fire Safety Order (FSO) generally replaces fire certification under the Fire Precautions Act 1971 with:

1 a general duty to take such fire precautions as may be reasonably required to ensure that premises are safe for the occupants (both employees and others), and
2 a general duty to carry out a risk assessment.

1.1.2 Legislation remaining in force

Legislation still in force includes:

▶ Building Regulations 2000 (England and Wales) as amended
▶ The Health and Safety (Safety Signs and Signals) Regulations 1996, and
▶ The Housing Act 2004.

1.2 The Regulatory Reform (Fire Safety) Order 2005 (SI 2005 No. 1541)

1.2.1 General

The Regulatory Reform (Fire Safety) Order is a 'deregulatory measure', which has consolidated much (over 100 pieces) of English and Welsh legislation that included fire safety requirements. The intention of the Order is to encourage fire prevention – prevention being better than cure – whilst retaining fire detection and alarm requirements. Under the Order, fire certificates are abolished and the 'responsible person' for each premises must carry out an assessment of the risks of fire and take action as necessary to reduce the risks of fire and also assess the fire detection and alarm provisions.

The Fire Safety Order requires:

▶ General fire precautions to be taken, article 8. General fire precautions are defined in article 4.
▶ Risk assessment, article 9.
▶ Provision of fire detection and fire-fighting equipment, article 13.

1.2.2 Duty to take general fire precautions

Article 8 states:

> **8. Duty to take general fire precautions**
>
> (1) The responsible person must –
>
>   (a) take such general fire precautions as will ensure, so far as is reasonably practicable, the safety of any of his employees; and
>
>   (b) in relation to relevant persons who are not his employees, take such general fire precautions as may reasonably be required in the circumstances of the case to ensure that the premises are safe.

Measures required to comply with this article include those necessary:

1 to reduce the risk of fire on the premises and the risk of the spread of fire on the premises;
2 to provide means of escape from the premises;
3 for securing that, at all material times, means of escape can be safely and effectively used;
4 to provide as may be appropriate means for fighting fires on the premises;
5 for detecting fire and giving warning in case of fire; and
6 to implement fire safety procedures, including:
    a the instruction and training of employees; and
    b limiting the effects of the fire.

Precautions to be taken will include as necessary the installation of:

1 emergency lighting
2 escape route signs
3 fire detectors and alarms.

IET publication *Electrician's Guide to Emergency Lighting* provides guidance on emergency lighting and escape signs.

1.2.3 Risk assessment

Article 9 is reproduced below:

9. Risk assessment

(1) The responsible person must make a suitable and sufficient assessment of the risks to which relevant persons are exposed for the purpose of identifying the general fire precautions he needs to take to comply with the requirements and prohibitions imposed on him by or under this Order.

(2) Where a dangerous substance is or is liable to be present in or on the premises, the risk assessment must include consideration of the matters set out in Part 1 of Schedule 1.

(3) Any such assessment must be reviewed by the responsible person regularly so as to keep it up to date and particularly if —

    (a) there is reason to suspect that it is no longer valid; or

    (b) there has been a significant change in the matters to which it relates including when the premises, special, technical and organisational measures, or organisation of the work undergo significant changes, extensions, or conversions,

and where changes to an assessment are required as a result of any such review, the responsible person must make them.

(4) The responsible person must not employ a young person unless he has, in relation to risks to young persons, made or reviewed an assessment in accordance with paragraphs (1) and (5).

(5) In making or reviewing the assessment, the responsible person who employs or is to employ a young person must take particular account of the matters set out in Part 2 of Schedule 1.

(6) As soon as practicable after the assessment is made or reviewed, the responsible person must record the information prescribed by paragraph (7) where —

    (a) he employs five or more employees;

    (b) a licence under an enactment is in force in relation to the premises; or

    (c) an alterations notice requiring this is in force in relation to the premises.

(7) The prescribed information is —

    (a) the significant findings of the assessment, including the measures which have been or will be taken by the responsible person pursuant to this Order; and

    (b) any group of persons identified by the assessment as being especially at risk.

(8) No new work activity involving a dangerous substance may commence unless —

    (a) the risk assessment has been made; and

    (b) the measures required by or under this Order have been implemented.

A five-step approach to risk assessment is recommended by the Department for Communities and Local Government:

Step 1 Identify fire hazards, sources of ignition, sources of fuel, sources of oxygen
Step 2 Identify people at risk
Step 3 Evaluate, remove, reduce and protect from risk: take preventative measures, take protective measures
Step 4 Record, plan, instruct, inform and train
Step 5 Review and revise

1.2.4 Guidance on risk assessment

Guidance is available from the Office of Public Sector Information including *Regulatory Reform (Fire Safety) Order 2005: A short guide to making your premises safe from fire.*

In addition to the short guide the following 12 sector-specific guides are available on the Department for Communities and Local Government website.

Entry level guide	A short guide to making your premises safe from fire
Guide 1	Offices and shops
Guide 2	Factories and warehouses
Guide 3	Sleeping accommodation
Guide 4	Residential care premises
Guide 5	Educational premises
Guide 6	Small and medium places of assembly
Guide 7	Large places of assembly
Guide 8	Theatres and cinemas
Guide 9	Outdoor events
Guide 10	Healthcare premises
Guide 11	Transport premises and facilities
Guide 12	Animal establishments
Supplementary Guide	Means of escape for disabled people

All the above may be downloaded from the Communities and Local Government website: www.communities.gov.uk/fire/firesafety/firesafetylaw/aboutguides.

1.2.5 Provision of fire detection and fire-fighting equipment

Article 13(1) is reproduced as follows:

13. Fire-fighting and fire detection

(1) Where necessary (whether due to the features of the premises, the activity carried on there, any hazard present or any other relevant circumstances) in order to safeguard the safety of relevant persons, the responsible person must ensure that –

(a) the premises are, to the extent that it is appropriate, equipped with appropriate fire-fighting equipment and with fire detectors and alarms; and

(b) any non-automatic fire-fighting equipment so provided is easily accessible, simple to use and indicated by signs.

1.2.6 Provision for Scotland and Northern Ireland

The Fire (Scotland) Act 2005 and the Fire Safety (Scotland) Regulations 2006 impose similar requirements – see in particular section 53 requiring measures to ensure safety of employees from fire including fire risk assessment and section 54 requiring measures to be taken to ensure safety of employees (including fire risk assessment).

It is understood that legislation will be introduced for Northern Ireland in a similar manner to Scotland – see the Fire and Rescue Services (Northern Ireland) Order 2006 and the Fire Safety (Northern Ireland) Regulations.

1.3 Building Regulations 2000 (England and Wales) (SI 2000 No. 2531 as amended)

The Building Regulations continue to apply to building work. The definition in regulation 3 of the Building Regulations is reproduced as follows.

3. Meaning of 'building work'

(1) In these Regulations 'building work' means –

(a) the erection or extension of a building;

(b) the provision or extension of a controlled service or fitting in or in connection with a building;

(c) the material alteration of a building, or a controlled service or fitting, as mentioned in paragraph (2);

(d) work required by regulation 6 (requirements relating to material change of use);

(e) the insertion of insulating material into the cavity wall of a building;

(f) work involving the underpinning of a building.

(g) work required by regulation 4A (requirements relating to thermal elements);

(h) work required by regulation 4B (requirements relating to a change of energy status);

(i) work required by regulation 17D (consequential improvements to energy performance).

(2) An alteration is material for the purposes of these Regulations if the work, or any part of it, would at any stage result –

(a) in a building or controlled service or fitting not complying with a relevant requirement where previously it did; or

(b) in a building or controlled service or fitting which before the work commenced did not comply with a relevant requirement, being more unsatisfactory in relation to such a requirement.

(3) In paragraph (2) 'relevant requirement' means any of the following applicable requirements of Schedule 1, namely –

Part A (structure)
paragraph B1 (means of warning and escape)
paragraph B3 (internal fire spread – structure)
paragraph B4 (external fire spread)
paragraph B5 (access and facilities for the fire service)
Part M (access to and use of buildings).

There are certain exemptions:

i   Section 4 of the Building Act 1984 provides exemptions for some buildings belonging to statutory undertakers.

ii Schedule 2 of the Building Regulations exempts various types of building according to their use or size, such as:

a   Buildings controlled under other legislation, for example: any building the construction of which is subject to the Explosives Acts 1875 and 1923. Any building (other than a building containing a dwelling or a building used for office or canteen accommodation) erected on a site in respect of which a licence under the Nuclear Installations Act 1965 is for the time being in force. A building included in the schedule of monuments maintained under section 1 of the Ancient Monuments and Archaeological Areas Act 1979.

b   Buildings not frequented by people.

iii Buildings belonging to, or that are occupied by, the Crown authorities are also exempt.

The Fire safety requirements are given in Part B of Schedule 1 to the Building Regulations, reproduced below.

Requirement	Limits on application
Means of warning and escape	
B1. The building shall be designed and constructed so that there are appropriate provisions for the early warning of fire, and appropriate means of escape in case of fire from the building to a place of safety outside the building capable of being safely and effectively used at all material times.	Requirement B1 does not apply to any prison provided under section 33 of the Prisons Act 1952 (power to provide prisons etc.).
Internal fire spread (linings)	
B2. (1) To inhibit the spread of fire within the building the internal linings shall:	
(a) adequately resist the spread of flame over their surfaces; and	
(b) have, if ignited, a rate of heat release which is reasonable in the circumstances.	
(2) In this paragraph 'internal linings' mean the materials lining any partition, wall, ceiling or other internal structure.	
Internal fire spread (structure)	
B3. (1) The building shall be designed and constructed so that, in the event of fire, its stability will be maintained for a reasonable period.	
(2) A wall common to two or more buildings shall be designed and constructed so that it adequately resists the spread of fire between those buildings. For the purposes of this sub-paragraph a house in a terrace and a semi-detached house are each to be treated as a separate building.	
(3) Where reasonably necessary to inhibit the spread of fire with the building, measures shall be taken, to an extent appropriate to the size and intended use of the building, comprising either or both of the following:	Requirement B3(3) does not apply to material alterations to any prison provided under section 33 of the Prisons Act 1952.
(a) sub-division of the building with fire-resisting construction;	
(b) installation of suitable automatic fire suppression systems.	

Requirement	Limits on application

(4) The building shall be designed and constructed so that the unseen spread of fire and smoke within concealed spaces in its structure and fabric is inhibited.

External fire spread

B4. (1) The external walls of the building shall adequately resist the spread of fire over the walls and from one building to another, having regard to the height, use and position of the building.

(2) The roof of the building shall adequately resist the spread of fire over the roof and from one building to another, having regard to the use and position of the building.

Access and facilities for the fire service

B5. (1) The building shall be designed and constructed so as to provide reasonable facilities to assist fire fighters in the protection of life.

(2) Reasonable provision shall be made within the site of the building to enable fire appliances to gain access to the building.

These safety requirements cover means of escape, fire alarms, fire spread, and access and facilities for the fire and rescue service. The guidance on ways of meeting the requirements was revised in 2006 and is given in:

- Approved Document B: Fire safety, Volume 1 – dwellinghouses (see Chapter 4) and
- Approved Document B: Fire safety, Volume 2 – buildings other than dwellinghouses (see Chapter 6).

It is important that there is consultation with local authorities at all stages. (Approval can be given by private sector approved inspectors instead of local authority building control.) When occupied, a building may need to meet the requirements of the Housing Act 2004. It should be noted that depending upon the level of risk once the building is occupied, the enforcing authority (for the Fire Safety Order and/or Housing Act) may require additional fire safety provisions, see section 1.5.

1.4   Building Regulations Scotland and Northern Ireland

1.4.1   Scotland

The Building (Scotland) Act 2003 (2003 asp 8)

This Act gives Scottish Ministers the power to make building regulations to secure the health, safety, welfare and convenience of persons in or about buildings and of others who may be affected by buildings or matters connected with buildings; to further the conservation of fuel and power; and to further the achievement of sustainable development.

The Building (Scotland) Regulations 2004

The Building (Scotland) Regulations 2004 are made under the powers of the Building (Scotland) Act 2003 and apply solely in Scotland.

The regulations apply to construction, alteration, conversion and demolition of buildings and also to the provision of services, fittings and equipment in or in connection with buildings except where they are specifically exempted from regulations 8 to 12 (Schedule 1 to regulation 3). Schedule 3 to regulation 5 lists work that must comply but does need a building warrant.

The regulations prescribe functional standards for buildings, which can be found in Schedule 5 to regulation 9. The regulations are amended periodically. However, it is the regulations in force at the time of the application that must be complied with.

Responsibility for compliance with building regulations

In Scotland, final responsibility for compliance with building regulations rests with the 'relevant person' (defined in section 17 of the Act) who will normally be the owner or developer of a building. However, any person carrying out work, including electrical work, has a duty to ensure their work complies with building regulations.

In Scotland there are no private approved inspectors: approval must be received from local authority building control verifiers. Further guidance is given in Chapter 10.

1.4.2   Northern Ireland

The Building Regulations (Northern Ireland) Order 1979 (as amended 1990) is the primary legislation for building regulations in Northern Ireland. It sets out the powers, duties, responsibilities and rights of the Department of Finance and Personnel, district councils and applicants in relation to building regulations matters, including giving the Department of Finance and Personnel responsibility for writing regulations in the form of secondary or 'subordinate' legislation. These Regulations currently are the Building Regulations (Northern Ireland) 2000 (as amended 2009) and the Building (Prescribed Fees) Regulations (Northern Ireland) 1997. Technical booklets are published by the Department of Finance and Personnel (www.dfpni.gov.uk) for most of the technical parts of the regulations. They provide construction methods that, if followed, will be deemed to satisfy the requirements of the Building Regulations.

There is no obligation to follow the methods or comply with the standards set out in the Technical booklets, however it will have to be demonstrated to the satisfaction of district councils that the requirements of the Building Regulations have been met.

As in Scotland there are no private approved inspectors, approval must be received from local authority building control officers.

More information is available on the internet at www.dfpni.gov.uk/index/law-and-regulation /building-regulations/br-technical-booklets.htm.

1.5   The Housing Act 2004

The Housing Act 2004 imposes particular requirements for houses in multiple occupation (HMOs) in England and Wales.

An HMO is defined as a building, or part of a building, such as a flat, that:

1   is occupied by more than one household and where more than one household shares – or lacks – an amenity, such as a bathroom, toilet or cooking facilities
2   is occupied by more than one household and which is a converted building – but not entirely self-contained flats (whether or not some amenities are shared or lacking)
3   is converted self-contained flats, but does not meet as a minimum standard the requirements of the 1991 Building Regulations, and at least one third of the flats are occupied under short tenancies.

The building is occupied by more than one household:

1   as their only or main residence
2   as a refuge for people escaping domestic violence
3   by students during term time
4   for other purposes prescribed by the government.

A household is:

1   a family (including a single person, couple and same sex couple)
2   a group sharing other relationships, such as fostering, carers and domestic staff.

An HMO must have a licence if:

1   it is three storeys or more (includes basements), and
2   it is occupied by five people or more.

Exemptions:

1   If the whole property is in self-contained flats.
2   If the basement is in commercial use and there are only two residential storeys above.

(Special exemptions are available to universities.)

The licence will specify the maximum number of people who may live in the HMO and will require:

1 a valid current gas safety certificate
2 proof that all electrical appliances and furniture are kept in a safe condition
3 proof that smoke alarms are correctly positioned and installed.

Part 1 of the Act requires a risk assessment procedure, the Housing Health and Safety Rating System (HHSRS). Technical guidance on the assessment of hazards from fire and preventive measures is contained in Housing Health and Safety Rating System Guidance (Version 2) issued in November 2004 (ISBN 1-855112-752-6). See also the guidance *Housing – Fire Safety*, which was developed by LACORS, the Chief Fire Officers Association (CFOA) and the Chartered Institute of Environmental Health (CIEH). The guidance provides advice on how to keep residential buildings safe from fire, explains how to carry out a fire risk assessment and includes a range of case studies.

Note: The Building Regulations generally only apply whenever 'building work' is undertaken, say the erection, extension or alteration of a building. They would not apply to existing premises. However, the Housing Act applies to all HMOs whenever they are so classified by virtue of the way in which they are occupied.

1.6   Parallel application of the Fire Safety Order and Building Regulations

In England and Wales most building work subject to the Building Regulations will also be subject to the Fire Safety Order once the work is complete and the building is occupied. The likely parallel application of the fire safety requirements of the Building Regulations and other legislation requires early and continuing consultation between the applicant and the administrating bodies.

The consultation procedures recommended by the Department for Communities and Local Government (see Chapter 2) are intended to prevent extra building work being necessary before a building can be occupied.

1.7   Building control, the fire and rescue authority and fire safety

Designers, developers and occupiers of buildings must consult the building control body with respect to fire safety.

The building control body (local authority or approved inspector) is responsible for checking compliance with the requirements of the Building Regulations. The Regulations are concerned with building work and with material changes of use (which may give rise to requirements for building work) and the requirements for fire safety will apply to most buildings.

The fire safety enforcing authority (for most premises the fire and rescue authority) is responsible for the implementation of the Fire Safety Order. The Order is concerned with the safety of people in or about the building and the enforcing authority must take into account hazards arising from the use of the building, particularly those with fire risks.

Building work that complies with the Building Regulations requirements for fire safety will normally be satisfactory when the building is occupied. However, there may be risks associated with the specific operations of the occupier that would not be covered by the Building Regulations and the designer must take these into account.

During the design and construction phase of a project the building control body will check on compliance with the requirements of the Building Regulations. The building control body consults the fire safety enforcing authority to ensure their requirements are met and that they are compatible with the Building Regulations requirements.

1.8 Local Acts

There may be provisions in other primary legislation, and in some places in Local Acts, which require consultation to take place between local authorities and fire safety enforcing authorities. Under regulation 13(6) of the Approved Inspectors Regulations, an approved inspector must consult the fire authority if a Local Act would have required the local authority to do so, had they been undertaking the building control function. The applicant should check with the building control body to see whether there are any Local Acts in force and if there are additional requirements.

1.9 Premises subject to licensing

Many premises are controlled through a licensing procedure in which fire safety matters are an important concern. Article 43 of the Fire Safety Order has the effect that those conditions of licence are overridden by any conditions of the Fire Safety Order (fire safety conditions cannot be included in a licence that could have been included in the Fire Safety Order).

> **43. Suspension of terms and conditions of licences dealing with same matters as this Order**
>
> (2) At any time when this Order applies in relation to the premises, any term, condition or restriction imposed by the licensing authority has no effect in so far as it relates to any matter in relation to which requirements or prohibitions are or could be imposed by or under this Order.

Before issuing a licence in relation to premises (premises such as nuclear installations, explosives factories and magazines or where hazardous substances are stored or used in quantities above specified limits) to which the Fire Safety Order applies, the licensing authority must first consult with the fire safety enforcing authority – see article 42.

1.10 Determination

1.10.1 The Fire Safety Order

If the responsible person, being under an obligation to do so, has failed to comply with any provision of the Fire Safety Order and they cannot agree with the enforcing authority on the measures which are necessary to remedy the failure, article 36 of the Order provides for a determination by the Secretary of State.

Article 36 may only be used to determine a dispute where the enforcing authority and the responsible person both agree that a failure to comply has occurred and agree to refer the question to the Secretary of State. Where they do not agree (or do not wish to use the determination process) then article 35 provides for appeals to a magistrates' court.

1.10.2 The Building Regulations

If the applicant and a building control body disagree about whether plans of proposed work are in conformity with Building Regulations, the applicant can seek a determination of the question from the Secretary of State (or in Wales, from the National Assembly) under either section 16(10) (local authorities) or section 50(2) (approved inspectors) of the Building Act 1984.

An application for a determination should be sent to the Department for Communities and Local Government (or in Wales, to the National Assembly for Wales) who will charge a fee of half the relevant plans charge subject to a minimum of £50 and maximum of £500. The applicant should explain why they consider the proposal does comply, accompanying the statement of case with relevant drawings and a copy of any rejection notice.

1.11 Terminology

The terms used in this Guide are explained below.

Applicant: The person responsible for obtaining approval. In practice this is often the agent of the owner or developer, such as the architect, and it is used in this Guide to include anyone acting for the applicant.

Approved inspector: A person with a notification in writing (that has not been withdrawn) from a body designated by the Secretary of State, that the person complies with the requirements of Statutory Instrument 2000 No. 2532 The Building (Approved Inspectors etc.) Regulations 2000.

(Note that builders and developers are required by law to obtain building control approval – an independent check that the Building Regulations have been complied with. There are two types of building control providers, namely the local authority and approved inspectors. Both building control bodies will charge for their services. Both may offer advice before work is started.

The Construction Industry Council (CIC) (www.cic.org.uk) maintains and operates the Approved Inspectors Register in accordance with the responsibilities entailed by CIC's appointment as a designated body with effect from 1 July 1996.)

Approved Inspectors Regulations: The Building (Approved Inspectors etc.) Regulations 2000 (SI 2000/2532, ISBN 0-11-099898-7, The Stationery Office Ltd).

Building control body: A term used to include both local authority building control and approved inspectors.

Building Regulations: The Building Regulations 2000 (SI 2000/2531, ISBN 0-11-099897-9, The Stationery Office Ltd). The Building (Scotland) Regulations 2004 (www.sbsa.gov.uk).

Consultation: Exchange of information and comment between organisations, such as building control bodies and fire and rescue authorities, which may include correspondence and face-to-face meetings with the fire and rescue authority – the authority discharging the functions of the fire and rescue authority under the Fire and Rescue Services Act 2004 in the area in which the premises are, or will be, situated. In most cases the enforcement responsibilities of fire and rescue authorities are delegated to the fire and rescue service.

Fire safety enforcing authority: The body responsible for enforcing the Fire Safety Order 2005. This is usually the fire and rescue authority.

Fire Safety Order: The Regulatory Reform (Fire Safety) Order 2005 (SI 2005/1541, ISBN 0-11-072945-5, The Stationery Office Ltd).

Local authority: The local authority empowered to carry out the building control function and to enforce Building Regulations in their area by virtue of section 91(2) of the Building Act 1984.

Responsible person: Article 3 of the Fire Safety Order defines responsible person for the purposes of the Order:

3. Meaning of 'responsible person'

In this Order 'responsible person' means –

(a) in relation to a workplace, the employer, if the workplace is to any extent under his control;

(b) in relation to any other premises not falling within paragraph (a) –

    (i) the person who has control of the premises (as occupier or otherwise) in connection with the carrying on by him of a trade, business or other undertaking (for profit or not); or

    (ii) the owner, where the person in control of the premises does not have control in connection with the carrying on by that person of a trade, business or other undertaking.

Article 5(1) imposes a burden on employers by placing a duty on them to ensure that the requirements of the Order and any regulations made under it are complied with in respect of the premises. This responsibility reflects the employer's responsibility under European Community law to ensure the safety of his employees. Article 5(2) imposes a similar duty on the responsible person in relation to non-workplaces. In such cases, the responsible person is under a duty to ensure that the requirements of the Order and any regulations made under it are complied with in respect of the premises, so far as they concern matters within his control.

Article 5(3) also imposes a similar duty on any person other than the responsible person who has to any extent control of the premises (the duty extending only so far as the extent of control). This would, for example, apply to a contractor who was responsible for maintaining a fire alarm system.

1.12   Web references

Communities and local government

http://www.communities.gov.uk/planningandbuilding

The Scottish Government is responsible for Scottish Building Regulations and preparing technical guidance on how to meet them. The Scottish Building Standards Agency (with former responsibility) was disbanded on 1 April 2008 and its functions were transferred back to the Scottish Government. The department is known as Scottish Building Standards.

The government also produces a guide to the procedures to be followed before building.

Technical and procedural handbooks and short guides are available on the SBS website: www.sbsa.gov.uk.

For Building Regulations within Northern Ireland

http://www.dfpni.gov.uk/index/law-and-regulation/building-regulations/br-legislation.htm

Regulatory Reform (Fire Safety) Order

http://www.opsi.gov.uk/si/si2005/20051541.htm

Building Regulations

http://www.planningportal.gov.uk

The building control approvals process

2

(Reference document: Department for Communities and Local Government publication, *Building Regulations and Fire Safety: Procedural Guidance.*)

2.1 Preliminary design stage: advice and consultation

The designer needs to determine:

1 Building Regulations requirements
2 Fire Safety Order requirements

and needs to:

3 consult with interested parties (including insurers)
4 decide on relevant technology to be used
5 decide on the category and coverage
6 conduct a design risk assessment.

(**Note:** Since the enactment of Regulatory Reform (Fire Safety) Order in 2006 there is no longer a requirement for a fire certificate.)

The designer will need to consult with the building control bodies at an early stage. (The building control body will liaise with the fire officer as necessary.)

When the building or part of the building is to be put to a use where the Fire Safety Order applies, the fire safety enforcing authority will have powers that may influence the design of the building.

The Fire and Rescue Services Act 2004 (section 6(2)(b)) requires every fire and rescue authority to give advice, when requested, with respect to buildings in the area of the fire and rescue authority on:

1 fire prevention
2 restricting the spread of fire
3 means of escape in case of fire.

Note: The fire and rescue authority – the authority discharging the functions of the fire and rescue authority under the Fire and Rescue Services Act 2004 in the area in which the premises are, or will be, situated. In most cases the enforcement responsibilities of fire and rescue authorities are delegated to the fire and rescue service.

The advice of the fire safety authority may include recommendations given in the best interest of fire safety, but which are not enforceable – that is, are advisory. The fire and rescue authority should separately identify such recommendations. Copies of their recommendations will be sent to the local authority.

2.2 Statutory consultation

The requirement to consult with the fire safety enforcing authority is statutory when the Fire Safety Order applies.

However, even when the Fire Safety Order does not apply, it is only sensible to consult with the fire safety enforcing authority, particularly if:

1 the fire safety measures are complex or unusual.
2 the building or development is such as to have an impact on the deployment or resources of the fire and rescue service.

Approved inspectors must consult with the fire safety enforcing authority before issuing an initial notice and before giving a plans certificate or final certificate to the local authority.

If the building control body is unhappy with the proposals, they will advise of the changes they require and will not consult with the fire safety enforcing authority until they believe the Building Regulations are likely to be complied with. Then they will send to the fire and rescue authority, drawings and supporting documents. The fire safety enforcing authority should respond within 15 working days with their comments. They should advise of recommendations that:

1 are advisory.
2 are necessary to meet fire safety legislation, other than the Building Regulations
3 will have to be complied with when (before) the building is occupied.

Note: The fire safety enforcing authority may make observations with respect to the Building Regulations.

2.3 Approval of plans

Compliance with plans approved by either the local authority or certified by an approved inspector can give protection from local authority notices requiring alterations.

When the building control body is satisfied that the plans comply with the requirements of the Building Regulations they should issue a notice or certificate approving the plans. They are required to issue a decision notice within the statutory time limit. They of course may reject the plans or issue an approval conditional upon further action.

If the consultation is statutory (where the Regulatory Reform (Fire Safety) Order applies), the building control body should send a copy of the notice or certificate to the fire authority.

2.4 Amended plans

When the design of a building is changed, the applicant must submit amended plans to the building control body. If this is following a statutory consultation and before or after the plans are approved, the building control body will, as a matter of good practice, consult again with the fire safety enforcing authority, as changes may affect fire safety.

Where an approved inspector has been appointed, changes in a building project may require the issue of an amendment notice to the local authority, altering the description of the work given in the original initial notice. The approved inspector will then need to formally consult with the fire safety enforcing authority again, under regulation 13 of the Approved Inspectors Regulations, if work introduced by the amendment notice concerns a building to be put to a use where the Fire Safety Order applies and is subject to Part B (of Schedule 1 to the Building Regulations) requirements.

2.5 Alterations notices

When (under article 29 of the Fire Safety Order) an 'alterations notice' has been served then the responsible person must notify the fire safety enforcing authority before making any changes to the premises, including any services, fittings or equipment in the premises or to the quantity of dangerous substances present in the premises. When notifying the fire safety enforcing authority the responsible person may be required to provide details of the changes proposed and a copy of the fire risk assessment.

Note: An alterations notice under article 29 of the Fire Safety Order may be served by the fire safety enforcing authority in relation to high-risk premises (or premises which would be high risk if any change were made to them). It alerts the fire safety enforcing authority to any potential problems and allows an intervention before changes are made which significantly increase the risk

Note: For 'responsible person' see section 1.11 in Chapter 1.

2.6 Construction

During construction the building control body may make inspections to check compliance with the Building Regulations, including Part B Fire safety. They will not necessarily check compliance with the requirements of the fire safety enforcing authority.

The Building Regulations have no requirements for fire risks during construction. However, the Construction (Health, Safety and Welfare) Regulations 1996 do have requirements.

Fire safety enforcing authority officers may, from time to time, inspect premises undergoing works of construction to check that the fire safety provisions and measures are adequate.

2.7 Completion

2.7.1 Building Regulations

On completion the applicant must notify the building control body not more than five working days after completion (regulation 15(4) of the Building Regulations).

> **15. Notice of commencement and completion of certain stages of work**
>
> (4) A person carrying out building work shall, not more than five days after that work has been completed, give the local authority notice to that effect.

If the building control body is satisfied that the work complies with the requirements of the Building Regulations, it should issue:

1. in the case of a local authority, a **completion certificate**, or
2. in the case of an approved inspector, after consultation with the fire safety enforcing authority, a **final certificate**.

(Note: There is no obligation on a local authority to issue a completion certificate unless one has been formally requested, or unless the building will be put to a use to which the Fire Safety Order applies.)

Approved inspectors must give the local authority a final certificate following completion of the work, if satisfied that the work complies with the applicable requirements of the Building Regulations. Approved inspectors are subject to a statutory time limit for the issue of a final certificate starting from occupation.

(The initial notice will cease to have effect after a grace period of four weeks (eight weeks for buildings where the Order does not apply) and building control will usually revert to the local authority, unless the local authority agrees to extend the period.)

Approved inspectors must consult with the fire authority before issuing a final certificate (regulation 13 of the Approved Inspectors Regulations). Local authorities would also be expected to consult at this stage. Regulation 13(3) is reproduced below.

> **13. Approved inspector's consultation with the fire authority**
>
> (3) Where this regulation applies, the approved inspector shall consult the fire authority –
>
>   (a) before or as soon as practicable after giving an initial notice in relation to the work;
>
>   (b) before or as soon as practicable after giving a relevant amendment notice in relation to the work;
>
>   (c) before giving a plans certificate (whether or not combined with an initial notice); and
>
>   (d) before giving a final certificate.

The building control body should send a copy of the completion/final certificate to the fire safety enforcement authority including where available a copy of the risk assessment and/or 'as built' record drawings.

2.8   Occupation

Before occupation the following statutory obligations must be met.

2.8.1   Building Regulations

If it is proposed to occupy a building where a Building Regulations completion or final certificate has not been issued by the building control body then the building control body must be notified as soon as possible of the date. Regulation 15 of the Building Regulations requires the applicant to give the local authority at least five working days' notice prior to occupation. In accordance with section 92 of the Building Act 1984 any such notice should be in writing.

2.8.2   The Regulatory Reform (Fire Safety) Order

Where the Order applies, the responsible person must complete the fire risk assessment and the provisions required to address the identified risks must be in place when the building is occupied.

The designer should have prepared a fire strategy as part of the design and approvals process and this, together with as-installed drawings, can form the basis of the responsible person's assessment.

2.8.3   Local Acts

There may be Local Acts in force, which include fire safety requirements. The local authority needs to be consulted, e.g. section 20 buildings (the London Building Acts 1939).

2.9   Web references

Building Regulations and Approved Documents

http://www.planningportal.gov.uk/england/professionals/en/1115314110382.htm

Building Regulations and Fire Safety: Procedural Guidance

http://www.planningportal.gov.uk/uploads/br/BR_PDF_B_PROCEDURAL.pdf

Regulatory Reform (Fire Safety) Order 2005

http://www.opsi.gov.uk/si/si2005/20051541.htm

Introduction to Approved Document B: Fire safety

3

3.1 Notification

In England and Wales all proposed work that is subject to the provisions of Part B, or of any other Part of Schedule 1 to the Building Regulations 2000, must be notified to the local authority. The work will be required to be inspected by the local authority building control department, or, at the election of the person carrying out the work, by an approved inspector.

Failure to comply with the requirements of Schedule 1 to the Building Regulations 2000 is a criminal offence. Local authorities also have powers to require the removal or alteration of work that does not comply with the requirements of Schedule 1.

The local authority's enforcement powers are suspended in a case where an approved inspector is carrying out building control. However, if a person carrying out building work fails to comply with instructions from an approved inspector to rectify non-compliant work, the approved inspector must cancel the 'initial notice', which brought the project under his or her supervision. Building control will then revert to the local authority.

Approved inspector: A person with a notification in writing (that has not been withdrawn) from a body designated by the Secretary of State, that the person complies with the requirements of Statutory Instrument 2000 No. 2532 The Building (Approved Inspectors etc.) Regulations 2000.

3.2 Summary

Approved Document B (Volumes 1 and 2) is one of a series of documents that have been approved and issued by the Secretary of State for the purpose of providing practical guidance with respect to the requirements of Schedule 1 to the Building Regulations 2000 for England and Wales (SI 2000 No. 2531).

Note 1: Approved Document B is available for download from www.communities.gov.uk.
Note 2: SI 2000 No. 2531 has been amended by the Building (Amendment) Regulations 2001 (SI 2001 No. 3335), the Building (Amendment) Regulations 2002 (SI 2002 No. 440), the Building (Amendment) (No. 2) Regulations 2002 (SI 2002 No. 2871), the Building (Amendment) Regulations 2003 (SI 2003 No. 2692), the Building (Amendment) Regulations 2004 (SI 2004 No. 1465) and the Building (Amendment) (No. 3) Regulations (SI 2004 No. 3210). These are accessible for download from www.opsi.gov.uk.

The Approved Documents provide guidance for some of the more common building constructions. The documents advise there may well be alternative ways of achieving compliance with the requirements and that there is no obligation to adopt any particular solution contained in an Approved Document. However, the guidance given is authoritative and the designer or installer adopting a different solution may have to demonstrate its compliance to the relevant authority.

The guidance contained in an Approved Document relates only to the particular requirements of the Regulations which that document addresses. The building work will also have to comply with the requirements of any other relevant paragraphs in Schedule 1 to the Regulations.

There are Approved Documents that give guidance on each of the other requirements in Schedule 1 and on regulation 7.

Approved Document B gives guidance on how to meet the requirements of Part B in terms of means of early warning and escape, minimising both internal and external fire spread and the provision of access and facilities for the fire service.

3.3 Supplementary guidance

The responsible government department (Department for Communities and Local Government) occasionally issues additional material to aid interpretation of the guidance contained in Approved Documents. This material may be conveyed in official letters to chief executives of local authorities and approved inspectors and/or posted on the websites accessed through www.odpm.gov.uk/buildingregs.

3.4 Changes to Approved Document B

In April 2007 a revised Approved Document B: Fire safety came into effect. As part of the revision the Approved Document was published in two volumes:

▶ Volume 1 – dwellinghouses
▶ Volume 2 – buildings other than dwellinghouses

This is to assist firms that specialise in domestic work in identifying the requirements for dwellings.

The changes for domestic fire alarms include the following:

1   The fire alarm system should comply with BS 5839-6:2004 *Fire detection and fire alarm systems for buildings. Code of practice for the design, installation and maintenance of fire detection and fire alarm systems in dwellings*. However, simple guidance has been retained which is reproduced in Chapter 4.
2   All fire alarms should have a standby power supply.
3   Where a dwellinghouse is extended, smoke alarms should be provided in the circulation spaces.

Other than doors between a dwellinghouse and an integral garage, fire doors need not be provided with self-closing devices.

3.5   Northern Ireland (Technical Booklet E)

See http://www.dfpni.gov.uk

The equivalent of Approved Document B in Northern Ireland is Technical Booklet E. This is similar to Approved Document B but requires additionally in dwellings smoke detectors in the principal habitable rooms (e.g. the lounge) and a heat detector in the kitchen. See Chapter 10 for Scotland.

Dwellings – Approved Document B, Volume 1

4

Note: Approved Document B applies in England and Wales. The equivalent document for Northern Ireland is Technical Booklet E which has similar requirements, but notably in dwellings requires additional detectors in the form of smoke alarms in the principal habitable room (e.g. the lounge) and a heat alarm in the kitchen. The Scottish equivalents are the Technical Handbooks, see Chapter 10.

4.1　Introduction

The guidance of the Department for Communities and Local Government on Part B of the Building Regulations in England and Wales has been provided in two volumes from April 2007:

▶ Approved Document B: Fire safety, Volume 1 – dwellinghouses
▶ Approved Document B: Fire safety, Volume 2 – buildings other than dwellinghouses

The volumes are stand-alone documents and similar in that they repeat common guidance.

This chapter concerns the guidance in Volume 1 – dwellinghouses. Volume 1 does not include recommendations for flats or buildings containing flats; this is given in Volume 2 – see Chapter 6.

> **Dwellinghouse** A unit of residential accommodation occupied (whether or not as a sole or main residence):
>
> a.　by a single person or by people living together as a family
> b.　by not more than six residents living together as a single household, including a household where care is provided for residents.
>
> Dwellinghouse does not include a flat or a building containing a flat.

The guidance in Approved Document B is structured as Part B of Schedule 1 to the Building Regulations as follows:

B1 To provide satisfactory means of giving an early warning of fire and a satisfactory standard of means of escape for persons in the event of fire in a building.

B2 To ensure that fire spread over the internal linings of buildings is inhibited.

B3 To provide for stability of buildings in the event of fire; with sufficient fire separation within buildings and between adjoining buildings; and to inhibit the unseen spread of fire and smoke in concealed spaces in buildings.

B4 To ensure that external walls and roofs have adequate resistance to the spread of fire over the external envelope, and that spread of fire from one building to another is restricted.

B5 To provide access for fire appliances to buildings and the provision of facilities in buildings to assist fire-fighters in the saving of life of people in and around buildings.

This chapter is concerned with part of B1, the provision of a satisfactory means of giving alarm of fire.

4.2   Means of warning and escape (B1)

The requirement:

Requirement	Limits on application
Means of warning and escape	
B1. The building shall be designed and constructed so that there are appropriate provisions for the early warning of fire, and appropriate means of escape in case of fire from the building to a place of safety outside the building capable of being safely and effectively used at all material times.	Requirement B1 does not apply to any prison provided under section 33 of the Prisons Act 1952 (power to provide prisons etc.).

The Approved Document advises:

> In the Secretary of State's view the requirement of B1 will be met if:
>
> a. there is sufficient means for giving early warning of fire for persons in the building;
>
> b. there are routes of sufficient number and capacity, which are suitably located to enable persons to escape to a place of safety in the event of fire; and
>
> c. the routes are sufficiently protected from the effects of fire, where necessary.

This publication is concerned with item a. above; b. and c. are construction features.

4.3   Design principles

The design of a fire system starts with a risk assessment for the particular dwelling and building for which the system is to be designed. This is to try to avoid not taking account of any unusual features of the particular building. Ideally the designer will attempt to identify all relevant factors including:

1   sources of fire (sources of ignition, fuel stores and sources, flammable materials)
2   escape routes
3   room layout (e.g. relationship between bedrooms and potential fire sources)
4   occupants (e.g. the elderly, the infirm, students)
5   building structure
6   fire spread through the building (effect on escape routes).

Guidance is given that it is reasonable to presume that a fire will not start at two places at once, and that the fire will initially only cause a hazard in one location. However, the fire may well spread to other parts of the building, perhaps as a result of the ignition of furnishings. The designer will be expected to take precautions to reduce the risk of fire starting and fire spread in areas common to more than one dwelling.

Good fundamental design of the building can limit the spread of fire. The designer of the fire alarm system will be aiming to ensure that the fire is detected early, and that escape facilitators (signs, lighting, closing doors) work under fire stress, particularly when smoke and noxious gases hinder escape.

The Approved Document is intended to provide for most people who will use the dwelling. However, there may be people whose specific needs are not met and this must be dealt with on a case-by-case basis, for example occupants who may be deaf – see Chapter 6, section 6.8.

4.4 Fire detection and fire alarm systems in dwellings

4.4.1 Introduction

Approved Document B Volume 1 requires suitable arrangements in all dwellinghouses to give early warning of fire.

Approved Document B specifies:

> **1.3** All new dwellinghouses should be provided with a fire detection and fire alarm system in accordance with the relevant recommendations of BS 5839-6:2004 [*Code of practice for the design, and installation and maintenance of fire detection and fire alarm systems in dwellings*] to at least a Grade D [see Figure 5.4 of Chapter 5] Category LD3 standard [see Figure 5.9].
>
> **1.4** The smoke and heat alarms should be mains operated and conform to BS 5446-1:2000 or BS 5446-2:2003 respectively. … They should have a standby power supply, such as a battery (either rechargeable or non-rechargeable) or capacitor.

4.4.2 Standard house smoke alarm installation requirements

So that electricians need not refer to BS 5839-6, Approved Document B Volume 1 gives specific guidance as to how the requirements of the Approved Document may be met in the most common situations.

Figures 4.1 and 4.2 show the typical minimum requirements for a standard house with the kitchen separated and not separated from the circulation space, respectively.

▼ **Figure 4.1** Minimum requirement for smoke alarms (standard house, no storey exceeds 200m^2 floor area)

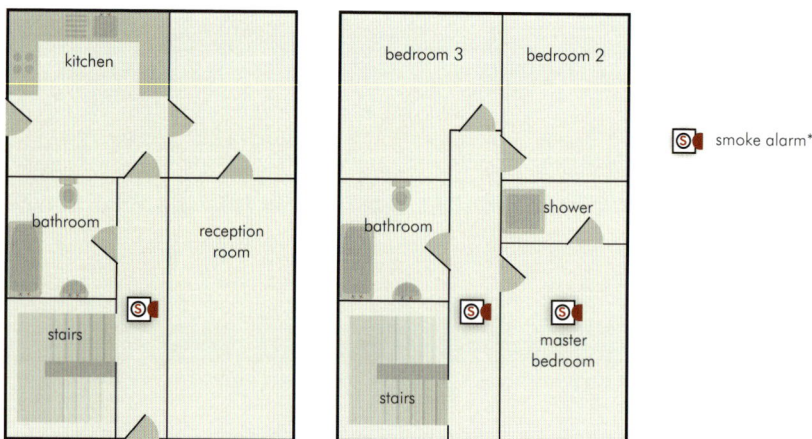

Notes to Figure 4.1:

*   Smoke alarm: a device containing within one housing all the components, except possibly the energy source, necessary for detecting smoke and for giving an audible alarm.

▶   Smoke alarms must be interlinked.

▶   Smoke alarms must normally be sited at least 300 mm from luminaires:

     a   in circulation spaces between kitchens and bedrooms

     b   within 7.5 m of every habitable room.

▶   The system is equivalent to BS 5839-6 grade D category LD3.

▼ **Figure 4.2**   Minimum requirements for smoke alarms (standard house, no storey exceeds 200m^2 floor area) with the kitchen not separated from the circulation space by a door

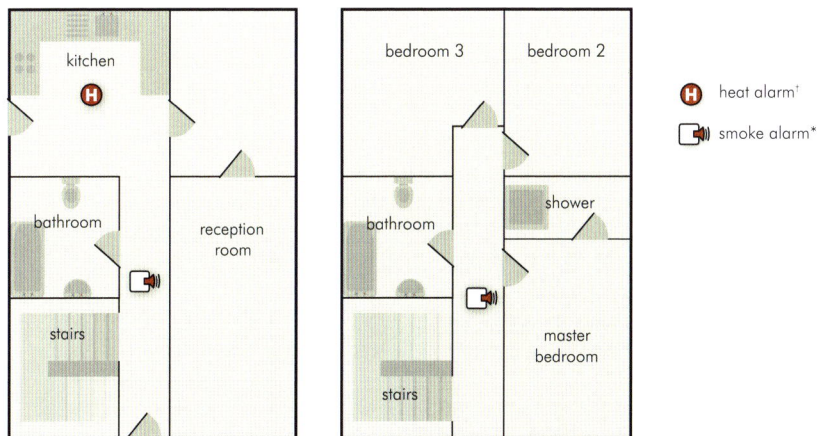

Notes:

*   Smoke alarm: a device containing within one housing all the components, except possibly the energy source, necessary for detecting smoke and for giving an audible alarm.

†   Heat alarm: a device containing within one housing all the components, except possibly the energy source, necessary for detecting heat and for giving an audible alarm.

▶   Alarms are to be interlinked.

1   The smoke and heat alarms should be mains-operated. They should have a secondary power supply such as a battery (either rechargeable or replaceable) or capacitor. More information on power supplies is given in Clause 15 of BS 5839-6:2004, see Section 5.7 of Chapter 5.

Note: Capacitor storage alarms have the advantage that batteries do not need replacing, bearing in mind that tenants may be reluctant to replace batteries at their own expense.

The smoke and heat alarms should conform to equipment standard BS 5446: *Fire detection and fire alarm devices for dwellings*, Part 1 *Specification for smoke alarms*; or Part 2 *Specification for heat alarms*.

2 The smoke alarms should normally be positioned in the circulation spaces between sleeping spaces and places where fires are most likely to start (e.g. between kitchens and bedrooms and between living rooms and bedrooms) to pick up smoke in the early stages.

3 In a house (including bungalows) there should be at least one smoke alarm on every storey.

4 Where the kitchen area is not separated from the stairway or circulation space by a door, there should be a compatible interlinked heat detector or heat alarm in the kitchen, in addition to whatever smoke alarms are needed in the circulation space(s).

5 Where more than one smoke alarm is installed they should be linked so that the detection of smoke or heat by one unit operates the alarm signal in all of them. The manufacturer's instructions about the maximum number of units that can be linked should be observed.

6 Smoke alarms/detectors should be sited so that:

  a there is a smoke alarm in the circulation space within 7.5 m of the door to every habitable room;

  b they are ceiling mounted and at least 300 mm from walls and light fittings (unless in the case of light fittings there is test evidence to prove that the proximity of the light fitting will not adversely affect the efficiency of the detector). Units designed for wall mounting may also be used provided that the units are above the level of doorways opening into the space, and they are fixed in accordance with manufacturers' instructions; and

  c the sensor in ceiling-mounted devices is between 25 mm and 600 mm (see note 2) below the ceiling (25–150 mm in the case of heat detectors or heat alarms).

Note 1: This guidance applies to ceilings that are predominantly flat and horizontal.
Note 2: A clearance of 600 mm would allow a lot of smoke to gather, so a clearance nearer 60 mm would be more appropriate.

7 It should be possible to reach the smoke alarms easily and safely to carry out routine maintenance, such as testing and cleaning. For this reason smoke alarms should not be fixed over a stair shaft or any other opening between floors.

8 Smoke alarms should not be fixed next to or directly above heaters or air-conditioning outlets. They should not be fixed in bathrooms, showers, cooking areas or garages, or any other place where steam, condensation or fumes could give false alarms.

9 Smoke alarms should not be fitted in places that get very hot (such as a boiler room), or very cold (such as an unheated porch). They should not be fixed to surfaces which are normally much warmer or colder than the rest of the space, because the temperature difference might create air currents which move smoke away from the unit.

Note from Approved Document B:

> BS 5446-1 covers smoke alarms based on ionisation chamber smoke detectors and optical (photo-electric) smoke detectors. The different types of detector respond differently to smoldering and fast flaming fires. Either type of detector is generally suitable. However, the choice of detector type should, if possible, take into account the type of fire that might be expected and the need to avoid false alarms. Optical detectors tend to be less affected by low levels of 'invisible' particles, such as fumes from kitchens that often cause false alarms. Accordingly they are more suitable than ionisation detectors for installation in circulation spaces adjacent to kitchens.

BS 5839-6 suggests that, in general, optical smoke alarms should be installed in circulation spaces such as hallways and landings. Optical detectors are also appropriate in areas in which a likely cause of fire is ignition of furniture or bedding by a cigarette and ionisation chamber-based smoke alarms may be the more appropriate type in rooms, such as the living room or dining room, where a fast-burning fire may present a greater danger to occupants than a smoldering fire.

4.4.3 Power supplies for Part B systems

1   The power supply for the fire alarm system should be taken from the dwelling's mains electricity supply. The mains supply to the smoke alarm(s) should comprise:

a   a single independent circuit at the dwelling's main distribution board (consumer unit); or

b   a single regularly used local lighting circuit. (This has the advantage that loss of supply will be rectified to restore the lighting.)

Most installations in dwellings, carried out in accordance with the 17th Edition of the IEE Wiring Regulations, will have a 30 mA RCD (residual current device) or RCBO (residual breaker with overcurret protection) protecting all circuits. In this circumstance it is arguably preferable for the supply for the fire alarm systems to be taken from a regularly used lighting circuit as operation of the RCD or RCBO will soon be known, see Figure 4.3. (For other than grade D systems, BS 5839-6 recommends avoiding the use of RCDs on fire system circuits. This will usually require clipping cables direct to the surface or enclosing in steel conduit or the like.)

The Approved Document requires a means of isolation for the fire system supply so that it can be isolated with the lights on. This will need to be labelled so that its purpose is clear.

Typical consumer unit arrangements are shown in Figures 4.3 and 4.4.

Note: See also Figure 5.1 and 5.2 of Chapter 5 for large houses. Where high reliability of the supply to the fire system (and security system) is required, for example for large houses and grade A and B systems, and RCD (or RCBO) protection to the circuit is to be omitted, cables will need to be installed in earthed steel conduit, see Figure 4.4.

▼ **Figure 4.3** Typical consumer unit arrangement

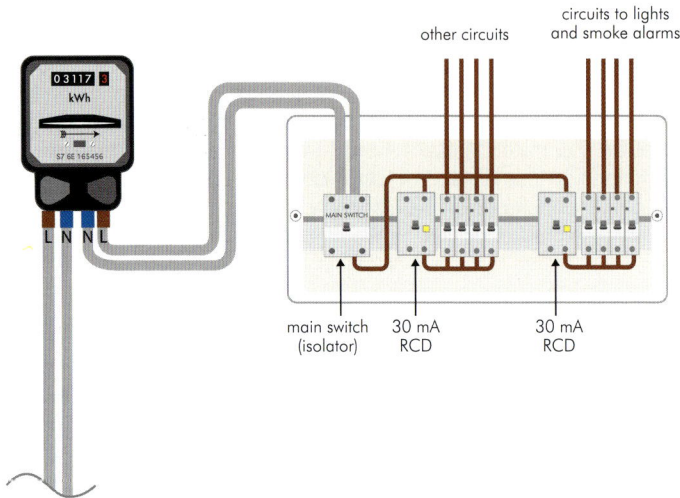

circuits to lights and smoke alarms

other circuits

03117 3
kWh

57 6E 16S456

L N N L

MAIN SWITCH

main switch (isolator)   30 mA RCD   30 mA RCD

▼ **Figure 4.4** Typical consumer unit arrangement – non-RCD protected circuits installed in earthed steel conduit

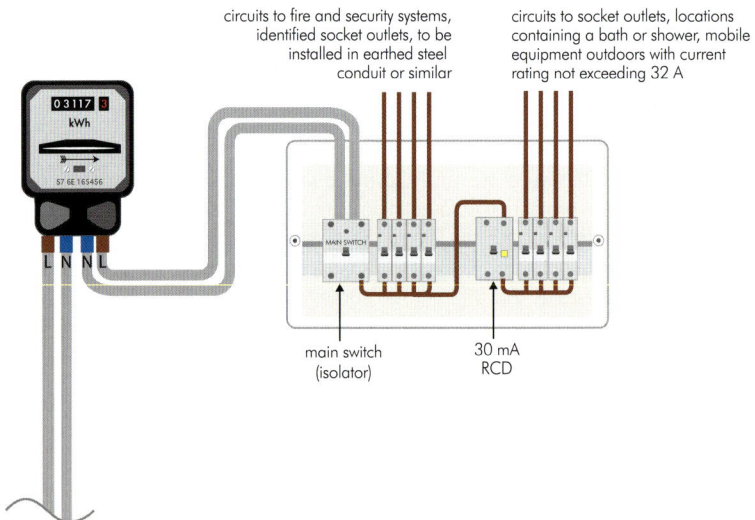

circuits to fire and security systems, identified socket outlets, to be installed in earthed steel conduit or similar

circuits to socket outlets, locations containing a bath or shower, mobile equipment outdoors with current rating not exceeding 32 A

03117 3
kWh

57 6E 16S456

L N N L

MAIN SWITCH

main switch (isolator)   30 mA RCD

2 Devices for monitoring the mains supply to the smoke alarm system may comprise audible or visible signals on each unit or on a dedicated mains monitor connected to the smoke alarm circuit. The design of any mains failure monitor should not reduce the reliability of the supply, and the alarm should be easily seen or heard by the occupants. If a continuous audible warning is used, it should be possible to silence it.

3 Any cable suitable for domestic wiring may be used for the power supply to and interconnection of smoke alarm units. It does not need any particular fire survival properties, except for large houses (see section 4.4.4) where BS 5839-6 grade A or grade B systems are required (see sections 5.3.1 and 5.3.2). Any conductors used for interconnecting alarms (signalling) should be identified from those supplying mains power, e.g. by colour.

Note: Smoke alarms may be interconnected using radio links, provided that this does not reduce the lifetime or duration of any standby power supply to below 72 hours.

4.4.4 Large houses

Note: A house is regarded as large if it has more than one storey and any of those storeys exceed 200 m^2.

Large house of two storeys

A large house of two storeys (excluding basement storey) should be fitted with a fire detection and alarm system grade B category LD3 as per BS 5839-6:2004.

For grade B see Table 5.1; for LD3 see Table 5.2.

Grade B: Fire detection and fire alarm system comprising fire detectors (other than smoke alarms and heat alarms), fire alarm sounders, and control and indicating equipment that either conforms to BS EN 54-2 (and power supply complying with BS EN 54-4) or to Annex C of Part 6 of BS 5839.

Power supply: A mains supply final circuit(s) to all parts of the fire alarm system dedicated solely to the fire alarm system, with a standby supply capable of automatically maintaining the system in normal operation (whilst giving an audible and visual indication of mains failure) for a period of 72 hours, after which sufficient capacity should remain to supply the maximum alarm load for at least 15 minutes; power supply equipment conforming to BS EN 54-4 *Fire detection and fire alarm systems*.

System LD3: A system incorporating detectors in all circulation spaces that form part of the escape routes from the dwelling. (In practice a grade A system is usually installed.)

Large house of three or more storeys

A large house of three or more storeys (excluding basement storey) should be fitted with a fire detection and alarm system grade A category LD2 as per BS 5839: Part 6: 2004 with detectors sited in accordance with the recommendations of BS 5839-1 for a category LD2 system; see Chapter 5 and Figure 4.5.

For grade A systems see Table 5.1; for LD2 category see Table 5.2.

Grade A system: A fire detection and fire alarm system, which incorporates control and indicating equipment conforming to BS EN 54-2, and power supply equipment conforming to BS EN 54-4, and which is designed and installed in accordance with all the recommendations of sections 1 to 4 inclusive of BS 5839-1:2002. A mains supply final circuit(s) to all parts of the fire alarm system dedicated solely to the fire alarm system, with a standby supply capable of automatically maintaining the system in normal operation (whilst giving an audible and visual indication of mains failure) for a period of 72 h, after which sufficient capacity should remain to supply the maximum alarm load for at least 15 minutes, power supply equipment conforming to BS EN 54-4 *Fire detection and fire alarm systems.*

LD2 category: A system incorporating detectors in all circulation spaces that form part of the escape routes from the dwelling, and in all rooms or areas that present a high fire risk to occupants.

Maximum alarm load: Maximum load imposed on a fire alarm system power supply under fire conditions, comprising the power required for simultaneous operation of all fire alarm devices, fire signals from all automatic fire detectors and manual call points in the building, any power drawn by other systems and equipment in the alarm condition and any power required for transmission of fire signals to an alarm receiving centre (if a facility for this is provided).

▼ **Figure 4.5**   Fire system for large house of three or more storeys (category LD2 grade A system)

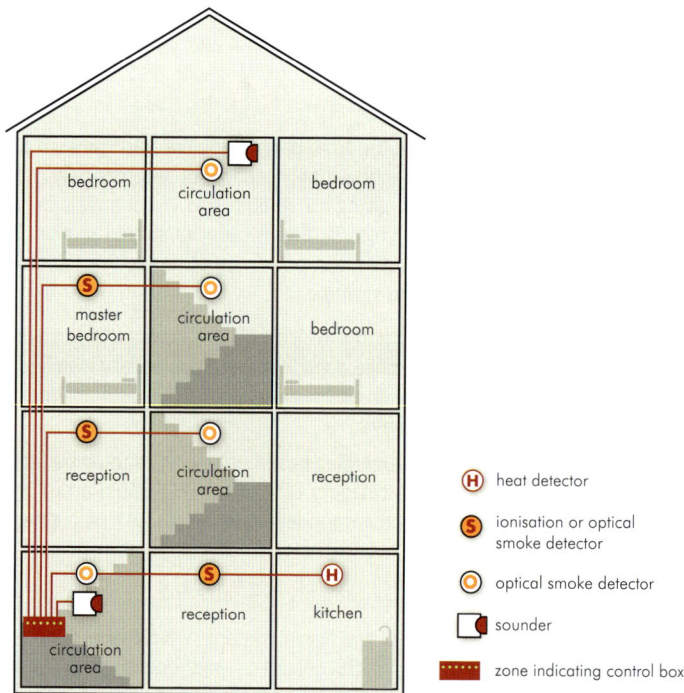

Symbol	Description
(H)	heat detector
(S)	ionisation or optical smoke detector
(O)	optical smoke detector
sounder	sounder
zone indicating control box	zone indicating control box

4.4.5 Extensions or alterations

Where habitable rooms are added above ground-floor level (e.g. bedrooms and loft conversions) then an automatic smoke detection and alarm system based on linked smoke and heat alarms should be installed throughout the dwelling to ensure that the occupants of the new rooms are warned of any fire (in the existing or new rooms) that may impede their escape. This is also necessary when extensions with rooms without direct escape to the outside are added to the ground floor (where a fire in the existing house might impede the escape of the occupants of the extension).

If the existing dwelling has no automatic fire detection and alarms, compliance with these requirements is most likely to require the installation of fire detection and alarms in the dwelling, though not necessarily in the extension.

In Figure 4.6 the extension has required a fire alarm and detection system to be installed in the existing dwelling to provide for safety in the extension.

▼ **Figure 4.6**   Fire system for extensions to a dwelling

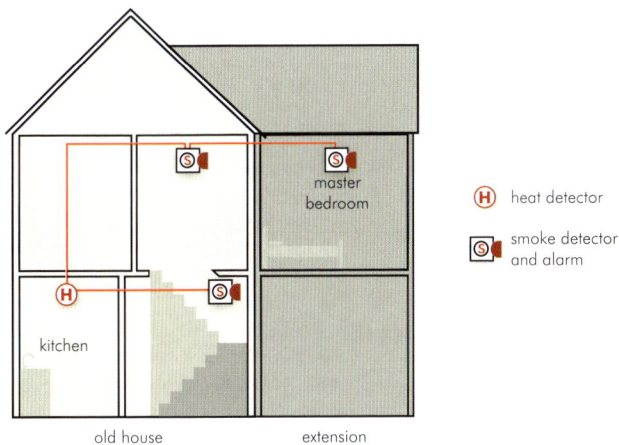

4.4.6 Flats

Flat: A separate and self-contained premises constructed or adapted for use for residential purposes and forming part of a building from some other part of which it is divided horizontally.

The requirements for flats are given in Approved Document B Volume 2; however, the requirements for within each flat are the same as for dwellings. Volume 2 includes the communal requirements for flats (or apartments) and this is discussed in section 6.5 of this Guide.

4.4.7 Sheltered housing

The detection equipment in a sheltered housing scheme with a warden or supervisor should have a connection to a central monitoring point (or alarm receiving station) so that either the warden or alarm receiving centre is aware that a fire has been detected in one of the dwellings, and can identify the dwelling concerned; see section 6.5.2 of this Guide.

4.4.8 Houses in multiple occupation

The guidance in Approved Document B (Volume 1) is applicable to houses in multiple occupation (HMOs), providing there are no more than six residents. (For HMOs with a greater number of residents, then additional precautions may be necessary.)

However the local authority has the authority to require additional fire safety provisions to those recommended in Approved Document B if appropriate. The Housing Act 2004 makes important requirements for HMOs. Architects and designers must always consult with the local authority to determine their requirements.

The Approved Document is intended to provide for most people who will use the dwelling. However there may be people whose specific needs are not met and this must be dealt with on a case-by-case basis, for example occupants who may be deaf – see section 6.8 of this Guide.

4.4.9 Radio links

Mains-powered smoke alarms may be interconnected by radio links, provided the lifetime or duration of the standby supply is not reduced to below 72 hours. If these conditions are met the smoke alarms may be connected to different circuits (as there is no need to isolate all the alarms to work on one).

4.5 Internal fire spread

For requirements on the control of fire spread, including sealing of penetrations by cables ducts and pipes, see section 8.5 of this Guide.

4.6 Inspection, testing and certification

4.6.1 Inspection and testing

The fire alarm electrical installation must comply with Part P of the Building Regulations. Test results may be included in the electrical installation certificate (and schedules) if carried out as part of the electrical installation; otherwise a minor electrical installation certificate shall be completed (for grade D systems). In all cases after successful testing and commissioning, a fire system certificate for design, installation and commissioning must be completed and given to the person ordering the work.

4.6.2 Commissioning

The system is inspected and set to work to confirm:

1 All manual call points and fire detectors work. Smoke detectors are tested to confirm that smoke initiates a fire alarm signal (e.g. by use of test aerosols as recommended by the manufacturer). Heat detectors are tested by means of a suitable heat source, as recommended by the manufacturer. (The heat source should not have the potential to ignite a fire and a live flame should not be used.)

2 Fire alarm warning devices (including any provided for deaf or hard-of-hearing people) work.

Where a fire alarm system is installed, an installation and commissioning certificate should be provided. The most appropriate are the standard forms of BS 5839-6 (see Chapter 5).

4.6.3 Hand-over

Inspection, test and commissioning certificates (and schedules), manufacturer's instructions and necessary guidance are to be handed over to the person ordering the work.

4.7 Maintenance

The building control body cannot practically make a requirement for maintenance as a condition of passing plans. However Approved Document B notes the importance of providing the occupants with information on the use of the equipment, and on its maintenance.

Note: BS 5839 Part 1 and Part 6 recommend that occupiers should receive the manufacturer's instructions concerning the operation and maintenance of the alarm system.

Dwellings – British Standard 5839-6

5

5.1 Introduction and Approved Document B

Approved Document B: Fire safety, Volume 1 – dwellinghouses guidance on the Building Regulations (see Chapter 4 of this Guide) specifies the installation of automatic fire detection in buildings within the scope of the Building Regulations – that is, in new dwellinghouses.

> **Dwellinghouse** A unit of residential accommodation occupied (whether or not as a sole or main residence):
>
> a. by a single person or by people living together as a family
> b. by not more than six residents living together as a single household, including a household where care is provided for residents.
>
> Dwellinghouse does not include a flat or a building containing a flat.

The detailed specifications of Approved Document B, Volume 1 are as follows:

a compliance with BS 5839-1 *Fire detection and fire alarm systems for buildings. Code of practice for system design, installation, commissioning and maintenance* to at least Category L3; or

b compliance with BS 5839-6 *Fire detection and fire alarm systems for buildings. Code of practice for the design, installation and maintenance of fire detection and fire alarm systems in dwellings* to at least Grade D, Type LD3; or

c provision of a suitable number of smoke alarms installed in accordance with paragraphs 1.4 to 1.22 of Part B (see Chapter 4 of this Guide).

The scope of BS 5839-6 is wider than that of Approved Document B in that it has recommendations for all dwellings including existing dwellings.

It is particularly useful in assessing the type of fire alarm systems to be installed in dwellings where the risk of fire and the risk of death are increased by the nature of the dwelling and the demography (age, lifestyle, etc.) of the inhabitants.

BS 5839-1 excludes dwellings from its scope; however, reference is made to 5839-1 in BS 5839-6 for grade A systems.

5.2　Scope of BS 5839-6 (BS 5839-6 Clause 1)

Part 6 of BS 5839 applies to all types of dwelling including: bungalows, multi-storey houses, individual apartments (flats and maisonettes), mobile homes, sheltered houses, housing providing 'NHS-supported living in the community', mansions, and houses divided into self-contained single-family dwelling units.

It does not apply to: hostels, caravans or boats (other than permanently moored boats used solely as residential premises), to the communal parts of purpose-built sheltered housing and blocks of apartments (flats or maisonettes), or to any premises used for purposes other than as a dwelling (e.g. small shops, factories or similar premises used solely as places of work).

The recommendations are primarily concerned with fire detection and fire alarm systems installed to protect life. However, recommendations are given for systems that are also intended to protect property.

5.3　Grades of system (BS 5839-6 Clause 7)

BS 5839-6 provides grades as well as categories of systems. The grade of a system determines the basic engineering design and the power supply, whilst the category specifies the coverage by detectors. For example, a Grade A system requires a mains power supply with a standby supply having a duration of at least 72 hours, designed in accordance with BS 5839-1 and conforming to BS EN 54-2 *Fire detection and fire alarm systems. Control and indicating equipment.* A Grade F installation is one with one or more battery-powered smoke alarms.

System grades relate to the engineering aspect of the fire detection alarm system. Higher grades tend to provide a greater level of control and monitoring or greater reliability and availability to perform in the event of a fire. The grade of the system that needs to be installed depends on the nature of the dwelling, the level of fire risk and the characteristics of the occupants.

Grade F systems (comprising battery-powered smoke alarms) are considered suitable for existing dwellings only. Grade F systems would not be appropriate in dwellings in which fire risk to occupants is high nor where there cannot be a reasonable certainty that the batteries would be replaced within a short time of a battery fault warning indication, nor for rented accommodation of two or more storeys. In these circumstances a system in which the supply is taken from the mains needs to be used, e.g. a Grade D. Table 5.1 summarises the fire system grades.

▼ **Table 5.1**   Summary of fire system grades

Grade	Description of system	Power supply
Grade A	A fire detection and fire alarm system, which incorporates control and indicating equipment conforming to BS EN 54-2, and power supply equipment conforming to BS EN 54-4, and which is designed and installed in accordance with all the recommendations of sections 1 to 4 inclusive of BS 5839-1:2002, except those in Clauses 16, 18, 20, 25.4e and 27, for which the corresponding clauses of Part 6 of BS 5839 should be substituted (i.e. 13, 14, 18, 15.2c and 21 respectively).	A mains supply final circuit(s) to all parts of the fire alarm system dedicated solely to the fire alarm system, with a standby supply capable of automatically maintaining the system in normal operation (whilst giving an audible and visual indication of mains failure) for a period of 72 h, after which sufficient capacity should remain to supply the maximum alarm load (see 3.22 of BS 5839-6) for at least 15 min; power supply equipment conforming to BS EN 54-4 *Fire detection and fire alarm systems. Power supply equipment.* No RCD unless required for electrical safety.
Grade B	A fire detection and fire alarm system comprising fire detectors (other than smoke alarms and heat alarms), fire alarm sounders, and control and indicating equipment that either conforms to BS EN 54-2 (and power supply complying with BS EN 54-4) or to Annex C of BS 5839 Part 6.	As for Grade A.
Grade C	A system of fire detectors and alarm sounders (which may be combined in the form of smoke alarms) connected to a common power supply, comprising the normal mains and a standby supply, with central control equipment.	As for Grade A but can be a common power supply with intruder alarm, comprising the normal mains and a standby supply, and reduced alarm time of 4 not 15 min.
Grade D	A system of one or more mains-powered smoke alarms, each with an integral standby supply. (The system may, in addition, incorporate one or more mains-powered heat alarms, each with an integral standby supply.) Smoke and heat detectors/alarms are interlinked (Clause 7.1).	Mains supply with an integral standby supply (battery or capacitor). If interconnected, on same circuit.
Grade E	A system of one or more mains-powered smoke alarms with no standby supply. (The system may, in addition, incorporate one or more heat alarms, with or without standby supplies.) Smoke and heat detectors/alarms are interlinked (Clause 7.1).	Mains supply, no standby supply, single independent circuit. No RCD unless required for electrical safety.
Grade F	A system of one or more battery-powered smoke alarms. (The system may, in addition, also incorporate one or more battery-powered heat alarms.) Smoke and heat detectors/alarms are interlinked (Clause 7.1).	Battery-powered.

5

5.3.1   Grade A system   (Clauses 15.2, 25.2h; BS 7671 Clause 522.6)

▼ **Figure 5.1**   Example of Grade A system

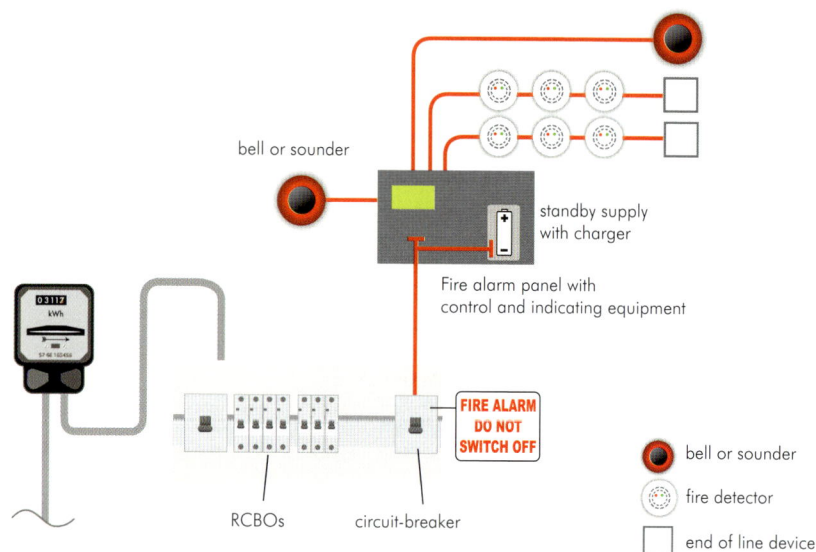

bell or sounder

standby supply with charger

Fire alarm panel with control and indicating equipment

FIRE ALARM DO NOT SWITCH OFF

RCBOs   circuit-breaker

● bell or sounder

⊙ fire detector

□ end of line device

Cables: To be fire resisting – see section 7.9.5 of this Guide. Standard fire resistant cables should:

1 meet the PH 30 classification when tested in accordance with BS EN 50200 (plus relevant test for fire and water), see note below; and
2 be selected from one of the following:
   a   mineral insulated copper sheathed cables, with an overall polymeric covering, conforming to BS EN 60702-1, with terminations conforming to BS EN 60702-2
   b   cables that conform to BS 7629 (fire resisting)
   c   cables that conform to BS 7846 (armoured fire resisting)
   d   cables rated at 300/500 V (or greater) that provide the same degree of safety to that afforded by compliance with BS 7629.

Note: Standard fire-resisting cables are required to meet the PH 30 classification and maintain circuit integrity when tested in accordance with BS EN 50200 and tests of simultaneously exposing a sample of the cable to flame at a temperature of 830 °C and mechanical shock for 15 min, then simultaneously exposing it to water spray and mechanical shock for a further 15 min.

In practice the installer needs confirmation from the cable supplier that cables selected meet the requirements of BS 5839-1 for standard fire-resistant cables.

Supply: To be separate with own isolator. Clause 25 of BS 5839-1 recommends that power supplies should not be protected by RCDs unless necessary for safety as in TT systems (i.e. where the electricity supplier does not provide an earth connection. The 17th Edition of the IEE Wiring Regulations (BS 7671: 2008) requires cables installed in dwellings in walls or partitions at a depth of less than 5 cm and installed in walls or partitions with metal parts to be protected by a 30 mA RCD unless the cables:

1   incorporate an earthed metal covering or
2   are installed in earthed steel conduit or trunking or similar.

Cables **2a**, and **2c** meet the requirement for an earthed metallic covering, but **2b** and **2d** would need to be installed in earthed steel conduit or similar or clipped direct to the surface.

Standby supply: Required with battery and charger.

5.3.2   Grade B system

▼ **Figure 5.2**   Example of Grade B system

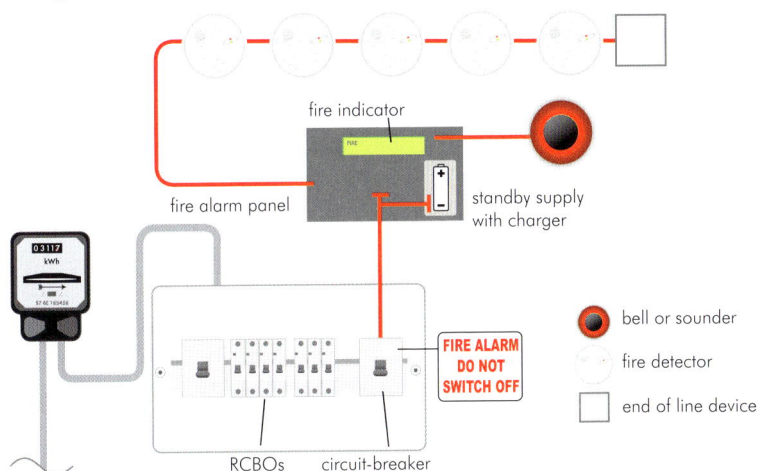

Grade B systems will have a much simpler panel to Grade A.

Cables: To be fire resisting see section 5.3.1 above.

Supply: Independent circuit not supplying other equipment.

Clause 15.3 of BS 5839-6 recommends that power supplies should not be protected by RCDs unless necessary for safety as in TT systems. As for A Grade systems, cables without an earthed metallic covering will generally require installation in steel conduit or similar or be clipped to the surface to comply with the 17th Edition of the IEE Wiring Regulations.

Standby supply: Required with battery and charger.

5.3.3   Grade C system

▼ **Figure 5.3**   Example of Grade C system

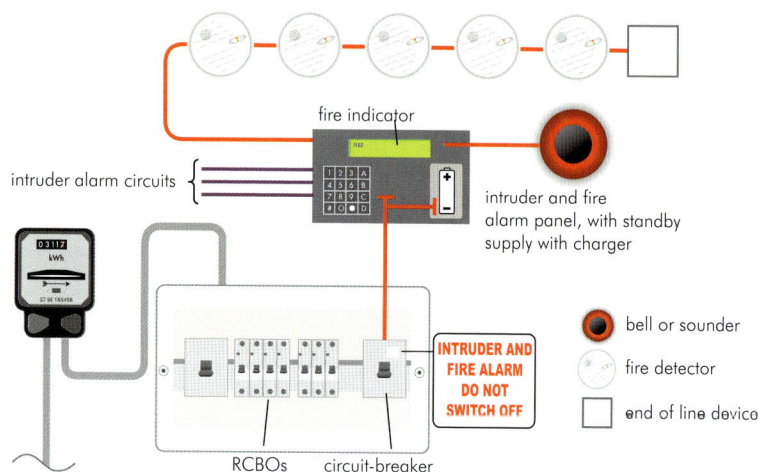

fire indicator

intruder alarm circuits

intruder and fire
alarm panel, with standby
supply with charger

INTRUDER AND
FIRE ALARM
DO NOT
SWITCH OFF

bell or sounder

fire detector

end of line device

RCBOs   circuit-breaker

Cables: Standard in accordance with BS 7671.

Supply: Independent circuit not supplying other equipment. Clause 15.3 of BS 5839-6 recommends that power supplies should not be protected by RCDs unless necessary for safety as in TT systems. As for A Grade systems, cables without an earthed metallic covering will generally require installation in steel conduit or similar or be clipped to the surface to comply with the 17th Edition of the IEE Wiring Regulations.

Standby supply: Required with battery and charger.

5.3.4 Grade D system

▼ **Figure 5.4** Example of Grade D system

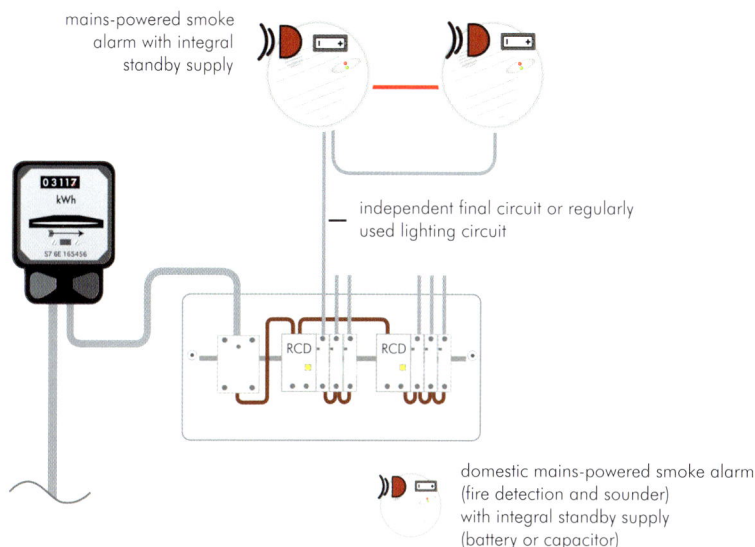

mains-powered smoke alarm with integral standby supply

independent final circuit or regularly used lighting circuit

domestic mains-powered smoke alarm (fire detection and sounder) with integral standby supply (battery or capacitor)

Cables: Standard in accordance with BS 7671.

Standby supply: Required either primary battery or more usually secondary battery or capacitor.

Supply: An independent circuit not supplying other equipment or regularly used lighting circuit should be used (Clause 15.5).

Interconnection: Required with standard cables per BS 7671 colour coded.

5.3.5 Grade E system

▼ **Figure 5.5** Example of Grade E system

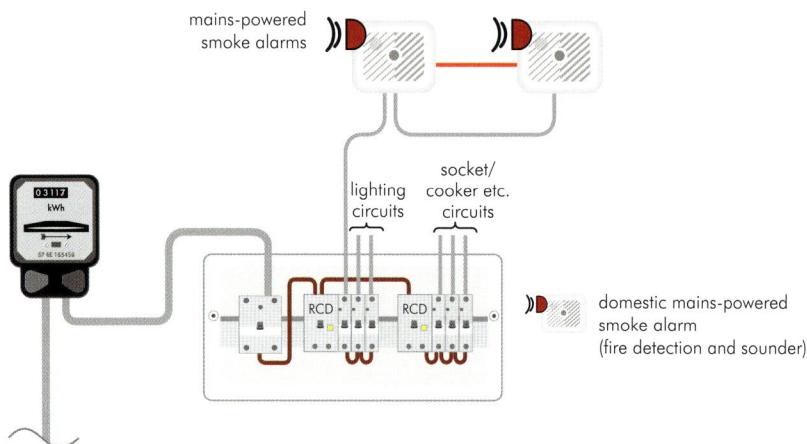

Cables: Standard in accordance with BS 7671.

Standby supply: Not required.

Supply: The fire alarm circuit should be a single independent circuit supplying no other equipment. The circuit should preferably not be protected by an RCD unless necessary for electrical safety as in TT systems. If an RCD is to be installed, Clause 15.6 recommends that:

1 the RCD should supply only the circuit supplying the alarms/detectors; or
2 the RCD should not supply any socket outlets or portable equipment.

To avoid the use of an RCD and comply with the 17th Edition of the IEE Wiring Regulations cables will need to incorporate an earthed metallic covering or be installed in steel conduit or similar or clipped direct.

Interconnection: Required with standard cables per BS 7671 colour coded.

5.3.6　Grade F system

▼ **Figure 5.6**　Grade F system

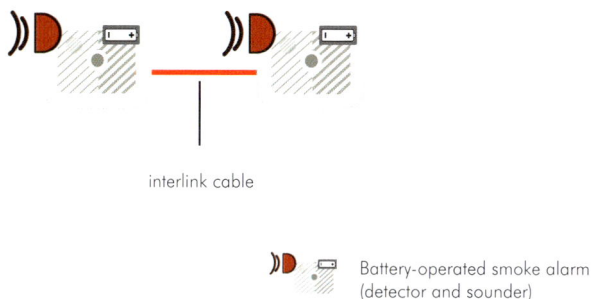

interlink cable

Battery-operated smoke alarm
(detector and sounder)

Supply: Battery with capacity to supply quiescent load for 1 year including weekly testing before low battery warning, then for 4-minute warning or further 30 days if no fire.

Interconnection: Recommended with standard cables per BS 7671.

5.4　Categories of system　(Clause 8)

The categories in BS 5839-6 are similar to those of BS 5839-1 except that the additional letter D is included and the subdivisions are specific to dwellings. The category identifies the level of protection provided:

▶ Letter L indicates protection of life
▶ Letter P indicates protection of property
▶ Letter D indicates a dwelling

Table 5.2 summarises the scope of the dwelling categories.

▼ **Table 5.2**   Categories for dwellings

Category	Description
Category LD	A fire detection and fire alarm system intended for the protection of life. Note 1: The designation 'LD' is used to distinguish these systems, which are intended only for dwellings, from Category L systems as defined in BS 5839-1, which are intended for the protection of life in any type of building.
Category LD1	A system installed throughout the dwelling, incorporating detectors in all circulation spaces that form part of the escape routes from the dwelling, and in all rooms and areas in which fire might start, other than toilets, bathrooms and shower rooms (see Figure 5.7).
Category LD2	A system incorporating detectors in all circulation spaces that form part of the escape routes from the dwelling, and in all rooms or areas that present a high fire risk to occupants (see Figure 5.8).
Category LD3	A system incorporating detectors in all circulation spaces that form part of the escape routes from the dwelling (see Figure 5.9).
Category PD	A fire detection and fire alarm system intended for the protection of property. Note 2: The designation 'PD' is used to distinguish these systems, which are intended only for dwellings, from Category P systems as defined in BS 5839-1, which are intended for the protection of property in any type of building.
Category PD1	A system installed throughout the dwelling, incorporating detectors in all rooms and areas in which fire might start, other than toilets, bathrooms and shower rooms.
Category PD2	A system incorporating detectors only in defined rooms or areas of the dwelling in which the risk of fire to property is judged to warrant their provision.

▼ **Figure 5.7**   Typical Category LD1

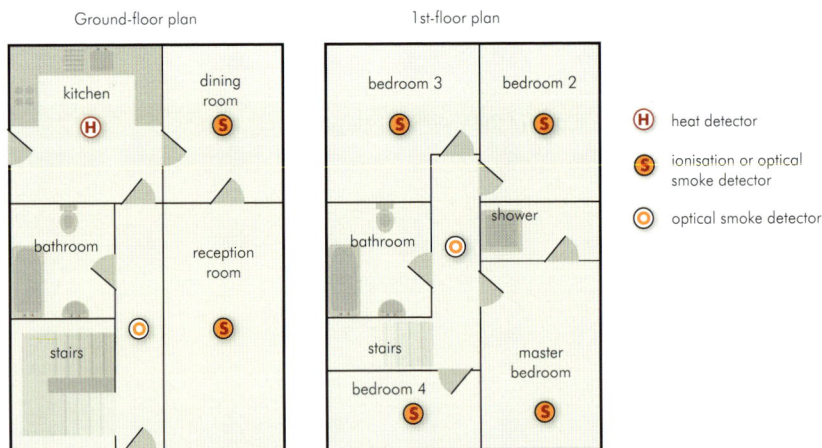

Ground-floor plan   1st-floor plan

kitchen — dining room — (H) — (S)
bathroom — reception room — stairs — (O) — (S)

bedroom 3 — bedroom 2 — (S) — (S)
bathroom — shower — (O)
stairs — master bedroom — bedroom 4 — (S) — (S)

(H) heat detector
(S) ionisation or optical smoke detector
(O) optical smoke detector

▼ Figure 5.8   Typical Category LD2

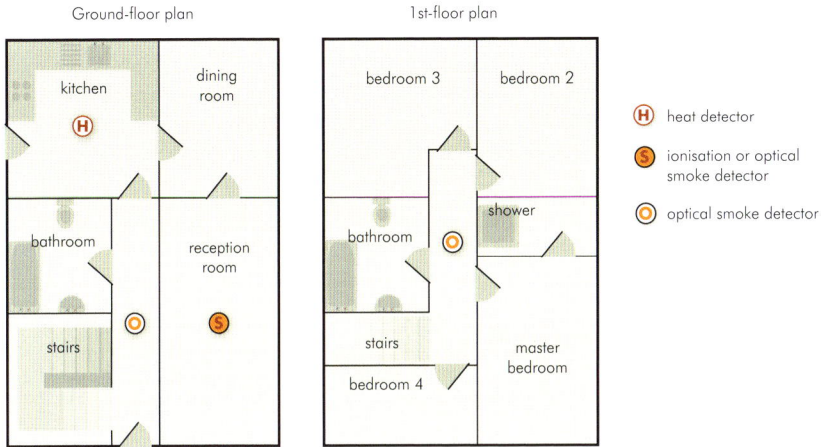

Ground-floor plan   1st-floor plan

Symbol	Description
(H)	heat detector
(S)	ionisation or optical smoke detector
(O)	optical smoke detector

▼ Figure 5.9   Typical Category LD3

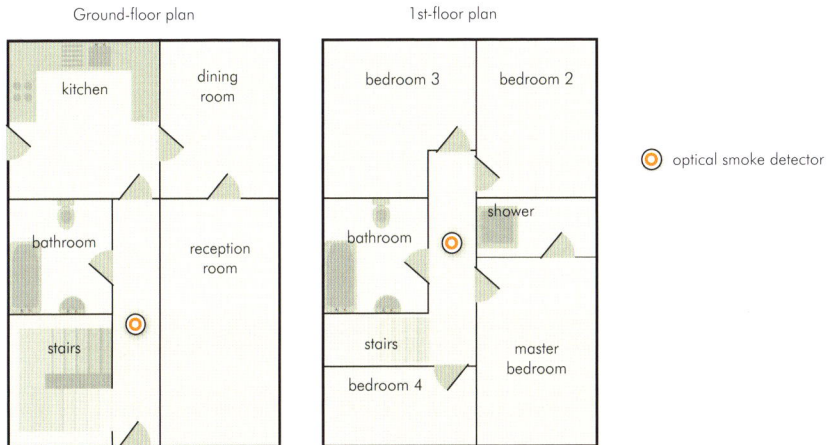

Ground-floor plan   1st-floor plan

Symbol	Description
(O)	optical smoke detector

5.5 Selection of grade and category (BS 5839-6 Clause 9)

5.5.1 Grade

System grades relate to the engineering aspect of the fire detection and alarm system with higher grades tending to provide greater level of control or greater reliability and availability to perform correctly in the event of a fire. They will have standby supplies. Lower grades such as Grade F are the least reliable because of the need for battery replacement.

5.5.2 Category

System categories relate to the level of protection afforded to the occupants; for example, an LD1 system installed throughout the dwelling and an LD3 in circulation spaces only.

5.5.3 Fire risk assessment

BS 5839-6 recommends that a fire risk assessment be carried out, particularly if a system is proposed to depart from the guidance given on grade and category selection.

A fire risk assessment will need to consider:

1 the dwelling – multiple occupations
2 the dwelling – design and escape routes
3 the dwelling – the integrity of each unit within the dwelling
4 the dwelling – the nature of the rooms
5 occupant characteristics – age, ability, social background
6 lifestyle factors
7 ignition sources.

Table 5.3 from Annex A of BS 5839-6 shows the relative frequency of fires in rooms within dwellings and is based on information provided by the Office of the Deputy Prime Minister (Department for Communities and Local Government).

▼ **Table 5.3**   Relative frequency of fire in rooms within dwellings

Room	Proportion of all domestic fires (%)
Kitchen	54
Bedroom, bedsitting room	12
Living room, dining room	12
Access area	6
Refuse area	3
Store room	2
Bathroom, cloakroom, WC	2
Roof space	1
Laundry	1
Airing cupboard, drying cupboard	1
Miscellaneous and unknown	7

BS 5839-6 provides two tables for the selection of grade and category. Table 5.4 is to be used for a specified dwelling in which the occupant characteristics are known or can be anticipated. Table 5.5 is described as providing recommendations for the minimum grade and category of a system that should be installed for property protection in a typical dwelling.

Notes to Table 5.4 (overleaf):

* Detectors sited in accordance with the recommendations of BS 5839-1 for the category system.

a In England and Wales, Approved Document B published by the Office of the Deputy Prime Minister.
In Scotland, the Technical Standards published by the Scottish Executive. In Northern Ireland, Technical Booklet E published by the Department of Finance and Personnel.

b Including dwellings with long-term lodgers, but not boarding houses, the latter of which are outside the scope of this part of BS 5839.

c Houses shared by no more than six persons, generally living in a similar manner to a single family (e.g. houses rented by a number of students).

d Heat detectors should be installed in every kitchen and the principal habitable room (see 3.28 of BS 5839-1). Where more than one room might be used as the principal habitable room, a heat detector should be installed in each of these rooms. The detector in the principal habitable room (but not the kitchen) may alternatively be a smoke or carbon monoxide fire detector. However, a heat detector is preferred in view of its lower potential for false alarms and the lesser need for maintenance.

e Grade E if there is any doubt regarding the ability of the occupier to replace batteries in battery-operated smoke alarms soon after a battery warning is given (see 9.1.1 of BS 5839-1) but Grade D, if, in addition, there is a significant likelihood of the electricity supply being disconnected because the occupier is unable to pay for supplies.

f Category LD2 if a risk assessment justifies the provision of additional detectors (see Clause 4).

g Detectors should be of a type and be so located as to compensate for the lower standard of structural fire precautions (for example, a smoke detector should be installed in the access room to a habitable inner room that has no door or window through which escape is possible). Further detectors might be necessary if a risk assessment justifies their provision. In some cases, a Category LD1 system might be necessary.

h The batteries in the smoke alarm(s) should have an anticipated life (taking into account monthly testing and fire alarm signals with an aggregate duration of 100 minutes per annum) of at least five years. Removal of batteries should necessitate the use of a tool.

i Further detectors might be necessary if a risk assessment justifies their provision.

j BS 5839-1 recommends that detectors are installed in escape routes and, generally, in rooms opening onto escape routes.
Notwithstanding the recommendations of BS 5839-1, detectors may be omitted from rooms opening directly onto staircase landings and opening onto escape corridors of 6 m or less in length.

k Other than houses with long-term lodgers and houses shared by no more than six persons, generally living in a similar manner to a single family (e.g. houses rented by a number of students).

l Detectors should be installed in communal circulation routes and within any circulation spaces in individual dwelling units comprising two or more rooms (e.g. in hallways and on staircase landings).

m The detectors in individual dwelling units may be incorporated within the system installed in communal areas.

n Category LD2 if a risk assessment justifies the provision of additional detectors (see Clause 4). For example, the conditions in a single-room bedsit might be such that the provision of a heat or smoke detector in the bedsit is justifiable.

Table 5.4 Minimum grade and category of fire detection and fire alarm system for protection of life in typical dwellings (Table 1 of BS 5839-6)

Class of dwelling	Minimum grade and category of system for installation in:					
	New or materially altered dwellings complying with the recommendations of BS 5588-1 or guidance that supports national building regulations[a]		Existing dwellings complying with the recommendations of BS 5588-1 or guidance that supports national building regulations[a]		Existing dwellings where structural fire precautions are of a lower standard than those recommended in BS 5588-1 or guidance that supports national building regulations[a]	
	Grade	Category	Grade	Category	Grade	Category
Single-family dwellings[b] and shared houses[c] with no floor greater than 200 m² in area						
Owner-occupied bungalow, flat or other single-storey unit	D	LD2d	Feh	LD3f	D	LD2B
Rented bungalow, flat or other single-storey unit	D	LD2d	Fe	LD3f	D	LD2B
Owner-occupied maisonette or owner-occupied two-storey house	D	LD2d	Fe	LD3f	D	LD2B
Rented maisonette or rented two-storey house	D	LD2d	D	LD3f	D	LD2B
Three-storey house	D	LD2d	D	LD3f	D	LD2B
Four- (or more) storey house	B	LD2d	D	LD2d*	B	LD2B
Single-family dwellings[b] and shared houses[c] with one or more floors greater than 200 m² in area						
Bungalow, flat or other single-storey unit	D	LD2d	D	LD3f	D	LD2B
Maisonette or two-storey house	B	LD2d	B	LD2di	B	LD2dg
Three- (or more) storey house*	A	LD2dij	A	LD2dij	A	LD2dij

▼ **Table 5.4** *continued*

Class of dwelling	Minimum grade and category of system for installation in:					
	New or materially altered dwellings complying with the recommendations of BS 5588-1 or guidance that supports national building regulations[a]		Existing dwellings complying with the recommendations of BS 5588-1 or guidance that supports national building regulations[a]		Existing dwellings where structural fire precautions are of a lower standard than those recommended in BS 5588-1 or guidance that supports national building regulations[a]	
	Grade	Category	Grade	Category	Grade	Category
Houses in multiple occupation[k] (HMOs)						
HMOs of one or two storeys with no floor greater than 200 m² in area	D	LD2[d]	D	LD3[lf]	D	LD2[lg]
Individual dwelling units, within the HMO, comprising two or more rooms	D[m]	LD2[d]	D[m]	LD3[n]	D[m]	LD2[g]
Communal areas of the HMO*	A	L2	A	L2	A	L2
Sheltered housing (individual dwelling units only)	C	LD2[d]	C	LD3[f]	C[q]	LD2[g]
Housing providing NHS-supported living in the community						
Dwellings of one, two or three storeys occupied by no more than six residents	C	LD1	C	LD1	C	LD1
Other dwellings	A	LD1	A	LD1	A	LD1

The notes to Table 5.4 are on pages 63 and 66.

Notes to Table 5.4 *continued*:

o Heat detectors should be installed in every communal kitchen. Heat or smoke detectors, as appropriate (taking into account the fire risk and the potential for false alarms), should be installed in every communal lounge.

p In sheltered housing, fire alarm signals from individual dwelling units should be relayed to the same location as alarm signals from any social alarm systems installed in the dwelling units. If there is an on-site warden, fire alarm signals should be investigated by the warden (e.g. using a two-way speech communication facility between the warden's facility and the dwelling) prior to summoning of the fire and rescue service. If no on-site warden is present, and fire alarm signals are transmitted to an alarm receiving centre that has a direct two-way speech facility for communication with the dwelling (of the type normally provided in social alarm systems), the alarm receiving centre should endeavour to determine, by use of this facility, whether the alarm signal is a false alarm, or has arisen as a result of a fire, before the fire and rescue service is summoned.

The recommendations assume that the on-site warden is readily available to respond immediately at all material times and that management standards at any alarm receiving centre comply with those recommended by BS 5979. If this cannot be ensured, consideration should be given to immediate summoning of fire and rescue services on each occasion that the fire alarm system operates.

q Many social alarm systems installed in sheltered housing have facilities for connection, monitoring and separate identification of signals from smoke alarms. If an existing social alarm system does not provide such a facility, it might be acceptable to install a Grade D system, provided that any fire signal is relayed to the warden's accommodation (see recommendation p).

▼ **Table 5.5** Minimum grade and category of fire detection and fire alarm system for protection of property in typical dwellings (Table 2 of BS 5839-6)

Class of dwelling	Grade	Category
Single-storey and two-storey dwellings	C	PD2
Other dwellings	A system conforming to the recommendations of BS 5839-1 for a category P1 system. This may be reduced to P2 if the risk to property from fire, and the maximum potential loss, do not warrant a Category P1 system.	

5.6   Types of fire detector

Fire detectors are designed to detect one or more of the four characteristics of fire:

1 smoke
2 heat
3 combustion gas (normally carbon monoxide)
4 flame.

Unfortunately all fire detectors will respond to some extent to phenomena not associated with a fire and this results in false alarms.

5.6.1   Smoke detectors

There are two classes of smoke detector commonly used:

1 Ionisation chamber smoke detectors. Ionisation smoke detectors are suitable for detecting fast flaming fires where there is little visible smoke.
2 Optical smoke detectors. Optical smoke detectors sense visible smoke particles and are effective for smouldering fires. They are the most suitable for general use.

BS 5839 advises that smoke detectors may be used generally other than in kitchens, bathrooms and shower rooms.

Smoke detectors installed within circulation areas, such as hallways, staircase landings and corridors, should generally be of the optical type.

Note from BS 5839-6: Custom and practice has been to use ionisation chamber smoke alarms in the above circulation areas. This practice is now deprecated in view of the greater potential for ionisation chamber smoke detectors to generate false alarms when exposed to fumes from kitchens, and in view of their poorer response to smouldering fires and smoke that has drifted some distance from its source.

5.6.2   Heat detectors   (Clauses 10.1.3 and 10.2)

Heat detectors respond to a fixed temperature being exceeded. They are suitable where high levels of dust and smoke make smoke detectors unsuitable. There are two main types of heat detector:

1 point detectors which respond to the temperature of the gases in the immediate vicinity of a single point
2 line detectors which respond to the temperature of the gases in the vicinity of a line (imaginary through the detector).

BS 5839 advises the following:

c) Heat detectors should not be used within circulation areas, such as hallways, staircase landings and corridors (i.e. heat detectors should not be used in any Category LD3 system).

Note 2: In the case of Grade A systems and systems complying with the recommendations of BS 5839-1 for a Category L3 system, it might be necessary to install detectors in rooms or areas opening onto escape routes; heat detectors may be used for this application.

d) In Category LD1, LD2 and PD systems, heat detectors may be installed within any room in a dwelling, unless it is necessary to give the earliest possible warning of a fire within the room (e.g. to protect sleeping occupants within the room or to protect high-value properties or their contents).

Note 3: A heat detector is unlikely to operate early enough to save the life of anyone asleep in the room in which it is installed. Moreover, a heat detector in the room of fire origin might not give sufficient warning for occupants elsewhere in the dwelling to escape safely if the door to that room is open.

Note 4: If a dwelling is protected by an automatic sprinkler system, and, on operation of any single sprinkler head, the fire detection and fire alarm system in the dwelling is activated (even if the mains power supply within the dwelling has failed), each sprinkler head may be regarded as a heat detector for the purpose of this part of BS 5839.

Heat detectors respond much more slowly than smoke detectors but are significantly less likely to give false alarms. They also generally require less maintenance than smoke detectors. They are not suitable for installation in circulation areas that form escape routes from a dwelling. The application for heat detectors depends on the category.

5.6.3 Carbon monoxide fire detectors   (Clauses 10.1.4 and 10.2)

Carbon monoxide fire detectors can be immune to certain environmental influences that can cause false alarms in other detectors such as dust, steam and cigarette smoke, whilst responding to most types of fire appreciably faster than most heat detectors.

Note: Carbon monoxide detectors to BS 7860 are intended only to detect carbon monoxide from faulty or inadequately ventilated gas appliances. As they are intended to give an alarm signal at much higher carbon monoxide levels than those to which carbon monoxide fire detectors are sensitive, they are not suitable for giving early warning of fire. Whilst detectors to BS EN 5029-1 are sensitive to lower carbon monoxide levels than those to BS 7860, they still respond too late to protect occupants against fire.

BS 5839-6 advises the following:

e) Carbon monoxide fire detectors, or multi-sensor detectors incorporating a carbon monoxide sensor, should not be used within dwellings, unless:

   i) the detectors are incorporated within a Grade A, B or C system and there is a high likelihood that the system will be subject to periodic maintenance by a competent person at periods not exceeding 12 months; or

   ii) a fault warning is given to indicate the need to replace the electrochemical cell of the detector before it reaches the end of its anticipated life.

f) Subject to compliance with **10.2**e) above, carbon monoxide fire detectors may be installed within the circulation areas of a dwelling in conjunction with an equal number of optical smoke detectors.

g) Subject to compliance with **10.2**e) carbon monoxide fire detectors may be installed in any rooms within a dwelling, other than kitchens.

Note 5: If, in any room of dwelling, a heat detector could provide adequate fire protection, a carbon monoxide detector may be used instead.

5.6.4 Flame detectors

Flame detectors are optical fire detection devices, which are able to detect infrared and/or ultraviolet radiation given off from a flaming fire.

Flame detectors are ineffective for slow smouldering fires where traditional smoke detectors would be more suitable. However, flame detectors will generally respond far quicker to rapidly developing fires such as combustible gases and liquids etc. It is most unlikely that flame detectors would be used to detect fires in dwellings.

5.6.5 Multi-sensor fire detectors

A multi-sensor fire detector conforming to BS EN 54-7 may be used when the use of a smoke detector would provide adequate fire protection. However, if a multi-sensor fire detector incorporates an ionisation chamber smoke sensor, it should only be used in circulation areas into which kitchens open if the system incorporates suitable measures to limit the potential for false alarms during cooking processes.

5.7 The electrical installation

The requirements for the electrical installation in terms of cable types and segregation as well as the main and standby supplies depend upon the system grade – see Table 5.6 which summarises the requirements.

5

▼ **Table 5.6** Electricity supply and wiring requirements of BS 5839-6 dwellings

(Clause 15)

Grade	Power supply (Clause 15)	Standby supply	Wiring (Clause 16)	Cable segregation
A	A mains supply final circuit(s) to all parts of the fire alarm system dedicated solely to the fire alarm system. Switches and protective device to be labelled 'FIRE ALARM: DO NOT SWITCH OFF'.	The standby supply should be capable of automatically maintaining the system in normal operation (whilst giving an audible and visual indication of mains failure) for a period of 72 h, after which sufficient capacity should remain to supply the maximum alarm load for at least 15 minutes. Secondary battery to have an automatic charger.	Cables and cable support systems (fixings, conduit, trunking) used for all parts of the critical signal paths and for the final circuit providing low-voltage mains supply to the system should be fire resisting. This can be achieved by ensuring either that cables are fire resisting or that cable and cable support systems (fixings, conduit, trunking) are fire resisting.*	Fire alarm cables to be segregated from cables of other services by, for example: 1 Not installing in the same conduit or trunking unless 'strong' compartments are used. 2 Use of separate sheathed multicore cables. 3 Physical separation. Fire cables to be identified by colour.
B	As for A.	As for A.	As for A.	No special requirements.
C	As A but may also supply an integral intruder alarm. The isolating protective device should be labelled 'FIRE/INTRUDER ALARM: DO NOT SWITCH OFF'.	As for A, the standby supply should be capable of automatically maintaining the system in normal operation for a period of 72 h (whilst giving the fault warnings), but after which sufficient capacity should remain to support the maximum alarm load for 4 min.	No special requirements.	No special requirements.
D	An independent circuit at the dwelling's main distribution board, in which case no other electrical equipment should be connected to this circuit (other than a dedicated monitoring device installed to indicate failure of the mains supply to the smoke alarms and any heat alarms); or a separately electrically protected, regularly used local lighting circuit.	As for C.	No special requirements.	No special requirements.

▼ Table 5.6 *continued*

Grade	Power supply (Clause 15)	Standby supply	Wiring (Clause 16)	Cable segregation
E	The mains supply to smoke alarm(s) and any heat alarms in a Grade E system should comprise a single independent circuit at the dwelling's main distribution board. No other electrical equipment should be connected to this circuit (other than a dedicated monitoring device installed to indicate failure of the mains supply to the smoke alarms and heat alarms).	None.	No special requirements.	No special requirements.
F	The batteries of smoke alarms and any heat alarms in Grade F systems should be capable of supplying the quiescent load of the smoke alarm or heat alarm, together with the additional load resulting from routine weekly testing, for at least one year before the battery fault warning is given. At the point at which the battery fault warning commences, the battery(ies) should have sufficient capacity to give a fire alarm signal for at least 4 min or, in the absence of a fire, a battery fault warning for at least 30 days.	None.	Not applicable.	Not applicable.

* Comprise either:
1   Mineral insulated copper sheathed cables, with an overall polymeric covering, conforming to BS EN 60702-1, with terminations conforming to BS EN 60702-2.
2   Cables that conform to BS 7629 specification for 300/500 V fire-resistant electric cables having low emission of smoke and corrosive gases when affected by fire.
3   Cables that conform to BS 7846. Electric cables – 600/1000 V armoured fire-resistant cables having thermosetting insulation and low emission of smoke and corrosive gases when affected by fire.
4   Cables rated at 300/500 V (or greater) that provide the same degree of safety to that afforded by compliance with BS 7629.

5.8   Model certificates

5.8.1   Model certificates for Grade A systems

The following model certificates are taken from Annex E of BS 5839-6.

E.1 Design certificate

Certificate of design for the fire alarm system at:

Address: ....

....

I/we being the person(s) responsible (as indicated by my/our signatures below) for the design of the fire alarm system, particulars of which are set out below, CERTIFY that the said design for which I/we have been responsible complies to the best of my/our knowledge and belief with the recommendations of Clauses 1 to 22 of BS 5839-6:2004 and the recommendations of Section 2 of BS 5839-1:2002 (as modified by the recommendations of BS 5839-6:2004) for the Grade A system described below, except for the variations, if any, stated in this certificate.

Name (in block letters): .... Position: ....

Signature: .... Date: ....

For and on behalf of: ....

Address: ....

....

.... Postcode: ....

The extent of liability of the signatory is limited to the system described below.

System category (see BS 5839-6:2004, Clause 8):

....

Variations from the recommendations of BS 5839-6 (including any variations from Section 2 of BS 5839-1:2002, other than those specifically recommended by BS 5839-6:2004):

....

....

....

....

....

Extent of system covered by this certificate: ....

....

Brief description of areas protected (not applicable for Category LD1 or PD1 systems):

....

....

....

....

....

Detector coverage is designed to satisfy the recommendations of BS 5839-1:2002 for a Category L1 or L2 system.

N/A ☐   L1 ☐   L2 ☐

Measures incorporated to limit false alarms. Account has been taken of the guidance contained in Section 3 of BS 5839-1:2002 and, more specifically (tick as appropriate):

☐ Account has been taken of reasonably foreseeable causes of unwanted alarms, particularly in the selection and siting of detectors.

☐ An appropriate analogue system has been specified.

☐ An appropriate multi-sensor system has been specified.

☐ A time-related system has been specified. Details: ....
....

☐ Automatic transmission of fire alarm signals to the fire and rescue service is delayed by .... mins by a delay in transmission of fire alarm signals to the alarm receiving centre/a delay, pending verification, before the alarm receiving centre summons the fire and rescue service.
(delete as appropriate)

☐ Appropriate guidance has been provided for the user to enable limitation of false alarms.

☐ Other measures as follows: ....
....

Installation and commissioning

It is strongly recommended that installation and commissioning be undertaken in accordance with the recommendations of Sections 4 and 5 of BS 5839-1:2002 respectively.

Soak test

☐ In accordance with the recommendations of **35.2.6** of BS 5839-1:2002, it is recommended that, following commissioning, a soak period of .... should follow. (Enter a period of not less than one week.)

☐ As the system incorporates no more than 50 automatic fire detectors, no soak test is necessary to satisfy the recommendations of BS 5839-1:2002.

Verification

Verification that the system complies with BS 5839-1:2002 should be carried out, on completion, in accordance with Clause **43** of BS 5839-1:2002:

Yes ☐   No ☐   To be decided by the ☐
purchaser or user

Maintenance

It is strongly recommended that, after completion, the system is maintained in accordance with Section 6 of BS 5839-1:2002.

> **This certificate may be required by an authority responsible for enforcement of fire safety legislation, such as the building control authority or housing authority. The recipient of this certificate might rely on the certificate as evidence of compliance with legislation. Liability could arise on the part of any organisation or person that issues a certificate without due care in ensuring its validity.**

E.2 Installation certificate

Certificate of installation for the fire alarm system at:

Address: ....

....

I/We being the person(s) responsible (as indicated by my/our signatures below) for the installation of the fire alarm system, particulars of which are set out below, CERTIFY that the said installation work for which I/we have been responsible complies to the best of my/our knowledge and belief with the specification described below and with the recommendations of Section 4 of BS 5839-1:2002, except for the variations, if any, stated in this certificate.

Name (in block letters): .... Position: ....

Signature: .... Date: ....

For and on behalf of: ....

Address: ....

....

.... Postcode: ....

The extent of liability of the signatory is limited to the system described below.

Extent of installation work covered by this certificate:

....

....

....

Specification against which system was installed:

....

....

....

Variations from the specification and/or Section 4 of BS 5839-1 (see BS 5839-1:2002, Clause 7):

....

....

....

Wiring has been tested in accordance with the recommendations of Clause **38** of BS 5839-1: 2002. Test results have been recorded and provided to:

....

Unless supplied by others, the 'as fitted' drawings have been supplied to the person responsible for commissioning the system.

Supplied to the person responsible [ ]   Supplied by others [ ]
for commissioning the system

This certificate may be required by an authority responsible for enforcement of fire safety legislation, such as the building control authority or housing authority. The recipient of this certificate might rely on the certificate as evidence of compliance with legislation. Liability could arise on the part of any organization or person that issues a certificate without due care in ensuring its validity.

E.3 Commissioning certificate

Certificate of commissioning for the fire alarm system at:

Address: ....

....

I/we being the person(s) responsible (as indicated by my/our signatures below) for the commissioning of the fire alarm system, particulars of which are set out below, CERTIFY that the said work for which I/we have been responsible complies to the best of my/our knowledge and belief with the recommendations of Clause **39** of BS 5839-1:2002, except for the variations, if any, stated in this certificate.

Name (in block letters): .... Position: ....

Signature: .... Date: ....

For and on behalf of: ....

Address: ....

....

.... Postcode: ....

The extent of liability of the signatory is limited to the systems described below.

Extent of system covered by this certificate: ....

....

....

Variations from the recommendations of Clause **39** of BS 5839-1:2002 (see BS 5839-1:2002, Clause 7): ....

....

....

- ☐ All equipment operates correctly.
- ☐ Installation work is, as far as can reasonably be ascertained, of an acceptable standard.
- ☐ The entire system has been inspected and tested in accordance with the recommendations of **39.2**c) of BS 5839-1: 2002.
- ☐ The system performs as required by the specification prepared by: ....
  a copy of which I/we have been given.
- ☐ Taking into account the guidance contained in Section 3 of BS 5839-1: 2002, I/we have not identified any obvious potential for an unacceptable rate of false alarms.
- ☐ The documentation described in Clause **40** of BS 5839-1:2002 has been provided to the user.

The following work should be completed before/after (delete as applicable) the system becomes operational:

....

....

The following potential causes of false alarms should be considered at the time of the next service visit:

....

....

Before the system becomes operational, it should be soak tested in accordance with the recommendations of **35.2.6** of BS 5839-1: 2002 for a period of:

....

(Enter a period of either one week, such period as required by the specification, or such period as recommended by the signatory to this certificate, whichever is the greatest, or delete if not applicable.)

> **This certificate may be required by an authority responsible for enforcement of fire safety legislation, such as the building control authority or housing authority. The recipient of this certificate might rely on the certificate as evidence of compliance with legislation. Liability could arise on the part of any organisation or person that issues a certificate without due care in ensuring its validity.**

E.4 Acceptance certificate

Certificate of acceptance for the fire alarm system at:

Address: ....

....

I/we being the competent person(s) responsible (as indicated by my/our signatures below) for the acceptance of the fire alarm system, particulars of which are set out below, ACCEPT the system on behalf of:

Name (in block letters): .... Position: ....

Signature: .... Date: ....

For and on behalf of: ....

Address: ....

....

.... Postcode:......................................

The extent of liability of the signatory is limited to the system described below.

Extent of system covered by this certificate: ....

....

....

☐   All installation work appears to be satisfactory.

☐   The system is capable of giving a fire alarm signal.

☐   The facility for remote transmission of alarms to an alarm receiving centre operates correctly. (Delete if not applicable.)

The following documents have been provided to the purchaser or user:

☐   'As fitted' drawings.

☐   Operating and maintenance instructions.

☐   Certificates of design, installation and commissioning.

☐   A log book.

☐   Sufficient representatives of the user have been properly instructed in the use of the system, including, at least, all means of triggering fire signals, silencing and resetting the system and avoidance of false alarms.

☐   All relevant tests, defined in the purchasing specification, have been witnessed. (Delete if not applicable.)

The following work is required before the system can be accepted:

....

....

....

....

....

....

5.8.2 Model certificate for Grades B, C, D, E and F systems

The following model certificate is taken from Annex F of BS 5839-6.

Certificate of design, installation and commissioning* of the fire detection and fire alarm system at:

Address: ....

....

....

....

It is certified that the fire detection and fire alarm system at the above address complies with the recommendations of BS 5839-6 for design, installation and commissioning of a Category ...., Grade .... system, other than in the respect of the following variations:*

....

....

....

....

Brief description of areas protected (only applicable to Category LD2 and PD2 systems).

....

....

....

....

The entire system has been tested for satisfactory operation in accordance with the recommendations of 23.3p) of BS 5839-6:2004*.

....

....

Instructions in accordance with the recommendations of Clause 24 of BS 5839-6:2004 have been supplied to:* ....

Signed: .... Date: ....

For and on behalf of: ....

* Where design, installation and commissioning are not all the responsibility of a single organisation or person, the relevant words should be deleted. The signatory of the certificate should sign only as confirmation that the work for which they have been responsible complies with the relevant recommendations of BS 5839-6:2004. A separate certificate(s) should then be issued for other work.

This certificate may be required by an authority responsible for enforcement of fire safety legislation, such as the building control authority or housing authority. The recipient of this certificate might rely on the certificate as evidence of compliance with legislation. Liability could arise on the part of any organisation or person that issues a certificate without due care in ensuring its validity.

Buildings other than dwellings – Approved Document B, Volume 2

6

Note 1: Buildings for some particular industrial and commercial activities presenting a special fire hazard, e.g. those involved with the sale of fuels, may require additional fire precautions to those detailed in Approved Document B.

Note 2: Approved Document B applies in England and Wales. The equivalent document for Northern Ireland is Technical Booklet E which has similar requirements. The Scottish equivalents are the Technical handbooks – see Chapter 10.

6.1 Introduction

The Department for Communities and Local Government's guidance on Part B (Fire safety) of the Building Regulations is now provided in two volumes:

▶ Volume 1 – dwellinghouses
▶ Volume 2 – buildings other than dwellinghouses

The volumes are stand-alone documents and similar in many sections as guidance common to both sections is repeated.

This chapter of the Guide concerns the guidance on fire alarms and fire detection systems given in Volume 2 – buildings other than dwellinghouses. Chapter 4 discusses the guidance on fire alarms and fire detection systems of Approved Document B, Volume 1 – dwellinghouses. The requirements for flats/apartments/maisonettes are included in Volume 2.

Note: Approved Document B is available for free download from the Communities and Local Government website: www.planningportal.gov.uk/england/professionals/buildingregs/technicalguidance.

As in Chapter 4 we consider here requirement B1 of Schedule I of the Building Regulations:

Requirement	Limits on application
Means of warning and escape	
B1. The building shall be designed and constructed so that there are appropriate provisions for the early warning of fire, and appropriate means of escape in case of fire from the building to a place of safety outside the building capable of being safely and effectively used at all material times.	Requirement B1 does not apply to any prison provided under section 33 of the Prisons Act 1952 (power to provide prisons etc.).

There is a brief introduction to requirements B2 to B5 in the Introduction to Chapter 4.

For large buildings 'fire engineering' consultants may be employed to look at fire risks and fire loads and to design specific fire protection schemes. Insurers will also be consulted.

6.2 Performance

(B1 of Approved Document B)

In the Secretary of State's view the performance requirements of B1 will be met if:

a. there are routes of sufficient number and capacity, which are suitably located to enable persons to escape to a place of safety in the event of fire;

b. the routes are sufficiently protected from the effects of fire by enclosure where necessary;

c. the routes are adequately lit;

d. the exits are suitably signed; and

e. there are appropriate facilities to either limit the ingress of smoke to the escape route(s) or to restrict the fire and remove smoke;

f. all to an extent necessary that is dependent on the use of the building, its size and height; and

g. there is sufficient means for giving early warning of fire for persons in the building.

6.3 Design principles (section B1.ii etc. of Approved Document B)

The design of the means of escape, and other fire safety measures such as the fire alarm system, should be based on an assessment of the risk to the occupants should a fire occur. The assessment should take into account the:

1 building structure
2 building use
3 materials stored
4 sources of fire
5 fire spread through the building, and
6 fire safety management proposed.

These assessments will be carried out by specialists.

The smoke and noxious gases are the prime danger, particularly in the early stages of a fire. Escape signs and escape routes may be obscured and occupants poisoned. Limiting the spread of smoke and fumes increases the chance of successful evacuation.

Guidance is given that it is reasonable to presume that a fire will not start at two places at once, and that the fire will initially only cause a hazard in one particular location. However, the fire may well spread to other parts of the building, perhaps as a result of the ignition of room contents.

After completion of the assessment, the responsible person (see section 6.12 below) will be expected to take precautions to reduce the risk of fire starting and reduce the risk of fire spreading to other parts of the building or to other flats/apartments in the same building.

Alarm system installers can bring this to the attention of the client.

The basic principles for the design of means of escape described in Approved Document B are as follows:

1 There should be alternative means of escape.
2 Where direct escape to a place of safety is not possible, it should be possible to reach a place of relative safety on a route to an exit, within a reasonable distance.
3 The ultimate place of safety is in the open air, clear of the effects of the fire (in a designated assembly area).

For compliance with the Building Regulations, the provision of the following is not acceptable as means of escape:

1 lifts (except for a suitably designed and installed evacuation lift that may be used for the evacuation of disabled people, in a fire);
2 portable ladders and throw-out ladders; and
3 manipulative apparatus and appliances, e.g. fold-down ladders and chutes.

6.4   Legislation   (B1.1 of Approved Document B V2)

The reader's attention is drawn to the legislation other than or additional to the Building Regulations, imposing requirements for means of escape in case of fire and other fire safety measures, with which the building must comply, and which will particularly come into force when the building is occupied, to provide for the safety of the persons using the building – see Chapter 1.

There are Acts and Regulations that impose fire safety requirements as a condition of a licence or registration. Whilst this other legislation is enforced by a number of different authorities, in the majority of cases the applicant and/or enforcing authority is required to consult the Fire Authority before a licence or registration is granted. The consultation may be carried out via the local authority. The fire and rescue authority are the primary body to be consulted and may make recommendations additional to the requirements of the Building Regulations. They will identify those recommendations that have statutory backing.

6.4.1   Standards

For more advanced fire safety design than that set out in Approved Document B, the fire safety design standard for other than dwellings is BS 9999: 2008 *Code of practice for fire safety in the design, management and use of buildings*, which has superseded BS 5588: *Fire precautions in the design, construction and use of buildings, Guide to fire safety codes of practice for particular premises/applications*, other than BS 5588-1.

BS 9999 catalogue descriptors are as follows:

> BS 9999 gives recommendations and guidance on the design, management and use of buildings to achieve acceptable levels of fire safety for all people in and around buildings.

> BS 9999 is applicable to the design of new buildings, and to alterations, extensions and changes of use of an existing building, with the exception of individual homes and with limited applicability in the case of certain specialist buildings. It also provides guidance on the ongoing management of fire safety in a building throughout the entire life cycle of the building, including guidance for designers to ensure that the overall design of a building assists and enhances the management of fire safety. It can be used as a tool for assessing existing buildings, although fundamental change in line with the guidelines might well be limited or not practicable.

This standard is considered in Chapter 9.

6.5 Fire alarm systems for flats

Flat: A separate and self-contained premises constructed or adapted for use for residential purposes and forming part of a building from some other part of which it is divided horizontally. (See Appendix E Definitions in Approved Document B, Volume 2.)

6.5.1 Introduction

The requirements for flats are given in Approved Document B, Volume 2 and BS 5588-1. The requirements for within each flat are similar to those for dwellings (e.g. Grade D, LD3) – see Chapter 4 of this Guide. Volume 2 includes the communal requirements for flats (or apartments) and is discussed below.

6.5.2 Sheltered housing

The detection equipment in a sheltered housing scheme with a warden or supervisor should have a connection to a central monitoring point (or central alarm relay station) so that the warden or person in charge, or a remote monitoring centre, is aware that a fire has been detected in one of the dwellings, and can identify the dwelling concerned. The requirements for communal areas are to be determined in accordance with the general requirements for buildings – that is, as appropriate to each particular building.

6.5.3 Houses in multiple occupation

(Houses in multiple occupation are defined at great length in section 254 of the Housing Act 2004. They are generally units of living accommodation occupied by persons who do not form a single household.)

The guidance in Approved Document B (Volume 1) is applicable to houses in multiple occupation (HMOs), providing there are no more than six residents. (For HMOs with a greater number of residents, then additional precautions may be necessary.)

However, the local authority has the authority to require additional fire safety provisions to those recommended in Approved Document B if appropriate. The Housing Act 2004 makes important changes to the legislation and designers must consult with the local authority to determine its requirements.

6.5.4 Responsibities of landlords and wardens etc.

Landlords and other responsible persons have legal duties (under the Regulatory Reform (Fire Safety) Order) to take measures to reduce the risk of fire, to reduce the spread of fire and to take measures to facilitate escape in communal areas (detection, alarms, escape routes).

This is an ongoing responsibility requiring risk assessments from time to time and continuing safety precautions including testing and maintenance – see sections 1.2.3 and 6.12 of this Guide.

6.6 Fire alarm systems – buildings other than flats

(paras 1.27 to 1.32 of Approved Document B V2)

Approved Document B, Volume 2 makes distinctions between buildings requiring:

1 electrically operated fire alarm systems initiated manually by people identifying a fire, and
2 electrically operated fire detection and fire alarm systems.

Approved Document B advises that whilst all buildings should have arrangements for detecting fire, in most buildings fires are detected by people, either through observation or smell, and therefore often nothing more will be needed with respect to detection. However, other than in the smallest of buildings with the few occupants within hailing distance of one another, the building should be provided with an electrically operated fire **warning** system with manual call points sited adjacent to exit doors and sufficient sounders to be clearly audible throughout the building. Insurers should be consulted for their requirements.

Volume 2 does not provide details of fire alarm systems except for flats (see section 6.9.5 of this Guide), and recommends that:

1 the fire alarm system should comply with BS 5839-1: *Fire detection and fire alarm systems. Code of practice for system design, installation, commissioning and maintenance* (see Chapter 7)
2 call points for electrical alarm systems should comply with Type A of BS EN 54 *Fire detection and fire alarm systems – Part 11: Manual call points.*

Volume 2 advises that if it is considered that people might not respond quickly to a fire warning, or where people are unfamiliar with the fire warning arrangements, a voice alarm system be considered so that both an alarm signal and verbal instructions would be given in the event of fire. If a voice alarm system is to be installed, it should comply with BS 5839-8 *Code of practice for the design, installation and servicing of voice alarm systems.*

Where there are large numbers of people, e.g. large shops, an initial general alarm may cause panic such as to inhibit pre-planned procedures for safe evacuation. The use of discreet sounders to alert staff may be necessary before evacuation procedures of the premises by sounders or messages broadcast over the public address system are initiated.

6.7 Automatic fire detection and fire alarm systems

(paras 1.35 to 1.38 of Approved Document B V2)

Automatic fire detection and alarms in accordance with BS 5839-1 should be provided in residential premises including 'institutional' premises (see Table 7.1 of Chapter 7).

The Approved Document advises that automatic fire detection systems are not normally needed in non-residential buildings including:

1 commercial (office and shop)
2 assembly and recreational
3 industrial
4 storage.

(This aligns with the guidance in BS 5839; see Table 7.3 of Chapter 7 of this Guide.)

However, a fire detection system may be needed, for example:

1 to compensate for some departure from the guidance;
2 as part of the operating system for some fire protection systems, such as pressure differential systems or automatic door releases;
3 where a fire could break out in an unoccupied part of the premises (e.g. a storage area or basement that is not visited on a regular basis, or a part of the building that has been temporarily vacated) and prejudice the means of escape from any occupied part(s) of the premises;
4 to meet insurers' requirements.

Fire detection and fire alarm systems should be in accordance with BS 5839-1, where general guidance on the standard of automatic fire detection that **may** need to be provided within a building can be found – see Chapter 7.

6.8 Warnings for deaf and hard-of-hearing people

(para 1.34 of Approved Document B V2)

Where it is expected that a number of deaf or hard-of-hearing people will use the building, where one or more such persons may be in relative isolation or where one or more such persons tend to move around the building to a significant extent, a suitable method of warning should be provided. In particular, all bedrooms and sanitary accommodation should have a visual and audible fire alarm signal.

Clause 18 of BS 5839-1:2002 *Fire detection and fire alarm systems for buildings* provides detailed guidance on the design and selection of fire alarm warnings for people with impaired hearing.

6.9   Particular premises

6.9.1   Hospitals (para 0.23 of Approved Document B V2)

NHS Estates has prepared a set of guidance documents on fire precautions in health care buildings, under the general title of *Firecode*, taking into account the particular characteristics of these buildings. These documents may also be used for non-NHS healthcare premises.

The design of fire safety in healthcare premises is covered by Health Technical Memorandum (HTM) 05-02 *Guidance in support of functional provisions for healthcare premises*. The approved document advises that where the guidance in that document is followed, Part B of the Building Regulations will be satisfied.

6.9.2   Unsupervised group homes (para 0.24 of Approved Document B V2)

Where a house of one or two storeys is converted for use as an unsupervised Group Home for not more than six mental health service users (mentally impaired or mentally ill people) it should be regarded as a Purpose Group 1(c) building if the means of escape are provided in accordance with HTM 05-02: *Guide to fire precautions in NHS housing in the community for mentally handicapped (or mentally ill) people*. Where the building is new, it may be more appropriate to regard it as being in Purpose Group 2(b).

6.9.3   Shopping complexes (para 0.25 of Approved Document B V2)

Approved Document B advises that a suitable approach to shopping complexes is given in Section 4 of BS 5588: Part 10: 1991 *Fire precautions in the design, construction and use of buildings, Code of practice for shopping complexes*.*

Note: BS 5588 Part 10* applies more restrictive provisions to units with only one exit in covered shopping complexes than given in BS 5588 Part 11 *Code of practice for shops, offices, industrial, storage and other similar buildings*.

6.9.4   Assembly buildings (para 0.26 of Approved Document B V2)

Approved Document B advises there are particular problems that arise when people are limited in their ability to escape by fixed seating. This may occur at sports events, theatres, lecture halls and conference centres etc. Guidance on this and other aspects of means of escape in assembly buildings is given in Sections 3 and 5 of BS 5588: Part 6: 1991 *Code of practice for places of assembly*† and the relevant recommendations of that code concerning means of escape in case of fire should be followed, in appropriate cases. The guidance given in the *Guide to fire precautions in existing places of entertainment and like premises* (HMSO) may also be followed. In the case of buildings to which the Safety of Sports Grounds Act 1975 applies, the guidance in the *Guide to safety at sports grounds* (HMSO) should be followed.

---

* Note that BS 5588-10 has now been superseded by BS 9999.
† Note that BS 5588-6 has been superseded by BS 9999.

6.9.5 Schools and other education buildings

(para. 0.27 of Approved Document B V2)

The design of fire safety in schools is covered by Building Bulletin (BB) 100 *Designing and managing against the risk of fire in shools*. Where the life safety guidance in that document is followed, Part B of the Building Regulations will be satisfied. Arson is a particular problem with schools, so means of starting fires must be minimised, e.g. keep bins away from buildings.

6.9.6 Flats

(para 1.2 to 1.5 of Approved Document B V2)

Approved Document B recommends a Grade D category LD3 system – see section 5.3.

6.10 Property protection

(para 0.23 of Approved Document B V2)

Building Regulations are intended to ensure that a reasonable standard of life safety is provided, in case of fire. The protection of property, including the building itself, may require additional measures.

Guidance is given in the Fire Protection Association's *Design guides for the fire protection of buildings*. Insurers should be consulted.

6.11 Supply of information

(para 0.12 of Approved Document B V2)

6.11.1 Introduction

In England and Wales regulation 16B of the Building Regulations 2000 (as amended in 2002) requires that sufficient information is provided for persons to operate, maintain and use the building in reasonable safety. This information is to assist the occupier/employer to meet their statutory duties under the Regulatory Reform (Fire Safety) Order. The exact amount of information and level of detail necessary will vary depending on the nature and complexity of the building's design. For small buildings, basic information on the location of fire protection measures may be all that is necessary. For larger buildings, a more detailed record of the fire safety strategy and procedures for operating and maintaining any fire protection measures of the building will be necessary. Appendix G of Approved Document B, Volume 1 gives advice on the sort of information that should be provided.

6.11.2 Operating and maintenance instructions
(Appendix G of Approved Document B V2)

Appendix G of Approved Document B, Volume 2 provides the following guidance information, which should be provided to assist the responsible persons to operate, maintain and use the building in reasonable safety and to assist the eventual owner, occupier and/or employer to meet their statutory duties under the Regulatory Reform (Fire Safety) Order.

1. This Appendix is only intended as a guide as to the kind of information that should be provided. For clarity the guidance is given in terms of simple and complex buildings, however the level of detail required will vary from building to building and should be considered on a case by case basis.

Simple buildings

2. For most buildings basic information on the location of fire protection measures may be all that is necessary. An as-built plan of the building should be provided showing:

a. escape routes;

b. compartmentation and separation (i.e. location of fire separating elements, including cavity barriers in walk-in spaces);

c. fire doors, self-closing fire doors, and other doors equipped with relevant hardware (e.g. panic locks);

d. locations of fire and/or smoke detector heads, alarm call-points, detection/alarm control boxes, alarm sounders, fire safety signage, emergency lighting, fire extinguishers, dry or wet risers and other fire fighting equipment, and location of hydrants outside the building;

e. any sprinkler system(s), including isolating valves and control equipment;

f. any smoke-control system(s) (or ventilation system with a smoke-control function), including mode of operation and control systems;

g. any high-risk areas (e.g. heating machinery);

h. specifications of any fire safety equipment provided, in particular any routine maintenance schedules; and

i. any assumptions in the design of the fire safety arrangements regarding the management of the building.

Complex buildings

3. For more complex buildings a more detailed record of the fire safety strategy and procedures for operating and maintaining any fire protection measures of the building will be necessary. Further guidance is available in BS 5588 Part 12 *Fire precautions in the design, construction and use of buildings: Managing fire safety* (Annex A Fire Safety Manual.)*

* Note that BS 5588-12 has been superseded by BS 9999.

These records should include:

a. The fire safety strategy, including all assumptions in the design of the fire safety systems (such as fire load). Any risk assessments or risk analysis.

b. All assumptions in the design of the fire safety arrangements regarding the management of the building.

c. Escape routes, escape strategy (e.g. simultaneous or phased) and muster points.

d. Details of all passive fire safety measures, including compartmentation (i.e. location of fire separating elements), cavity barriers, fire doors, self-closing fire doors and other doors equipped with relevant hardware (e.g. electronic security locks), duct dampers, and fire shutters.

e. Fire detector heads, smoke detector heads, alarm call-points, detection/ alarm control boxes, alarm sounders, emergency communications systems, CCTV, fire safety signage, emergency lighting, fire extinguishers, dry or wet risers and other fire fighting equipment, other interior facilities for the fire and rescue service, emergency control rooms, location of hydrants outside the building, other exterior facilities for the fire and rescue service.

f. Details of all active fire safety measures, including:
   - Sprinkler system(s) design, including isolating valves and control equipment; and
   - Smoke-control system(s) (or HVAC system with a smoke-control function) design, including mode of operation and control systems.

g. Any high-risk areas (e.g. heating machinery) and particular hazards.

h. As-built plans of the building showing the locations of the above.

i. Specifications of any fire safety equipment provided, including operational details, operators manuals, software, system zoning, and routine inspection, testing and maintenance schedules. Records of any acceptance or commissioning tests.

j. Any provision incorporated into the building to facilitate the evacuation of disabled people.

k. Any other details appropriate for the specific building.

6.12   Risk assessments and responsible person

The Regulatory Reform (Fire Safety) Order (commonly known as the Fire Safety Order), which came into force on 1 October 2006, applies generally including to the common parts of blocks of flats as well as commercial and industrial premises and requires responsible persons to carry out a fire risk assessment – see Communities and Local Government website: www.opsi.gov.uk/si/si2005/20051541.htm.

Note: Common parts of flats are not within the scope of legislation in Scotland and Northern Ireland. A fire warning system is not normally needed in the common parts of flats (in England, as well as Scotland and Northern Ireland).

6.12.1 Meaning of responsible person

The definition of responsible person given in the Order is reproduced below.

3. In this Order 'responsible person' means —

(a) in relation to a workplace, the employer, if the workplace is to any extent under his control;

(b) in relation to any premises not falling within paragraph (a) —

  (i) the person who has control of the premises (as occupier or otherwise) in connection with the carrying on by him of a trade, business or other undertaking (for profit or not); or

  (ii) the owner, where the person in control of the premises does not have control in connection with the carrying on by that person of a trade, business or other undertaking.

6.12.2 Risk assessment

Article 9(1) of the Regulatory Reform (Fire Safety) Order requires:

The responsible person must make a suitable and sufficient assessment of the risks to which relevant persons are exposed for the purpose of identifying the general fire precautions he needs to take to comply with the requirements and prohibitions imposed on him by or under this Order.

See sections 1.2.3 and 1.2.4 of Chapter 1 of this Guide for further guidance.

6.12.3 Fire precautions

General fire precautions given in article 4(1) of the Order are:

> (1) (a) measures to reduce the risk of fire on the premises and the risk of the spread of fire on the premises;
>
> (b) measures in relation to the means of escape from the premises;
> (c) measures for securing that, at all material times, the means of escape can be safely and effectively used;
> (d) measures in relation to the means for fighting fires on the premises;
> (e) measures in relation to the means for detecting fire on the premises and giving warning in case of fire on the premises; and
> (f) measures in relation to the arrangements for action to be taken in the event of fire on the premises, including –
> (i) measures relating to the instruction and training of employees; and
> (ii) measures to mitigate the effects of the fire.

6.13 Cables (protected power circuits)

(para 5.38 of Approved Document B V2)

Where 'it is critical for electrical circuits to continue to function during a fire' the Approved Document requires appropriate selection of cables, cable routes, cable fixing and refers to the PH 30 classification when tested to BS EN 50200. Meeting the requirements of BS 5839-1 for standard fire-resisting cables and their fixing plus appropriate selection of cable route would meet these requirements – see section 7.9.5 of Chapter 7 of this Guide.

Further advice is given that in large or complex buildings, fire protection systems may need to operate for extended periods and that then guidance should be sought from the recommendations of:

▶ BS 5839-1 *Fire detection and alarm systems for buildings* (see Chapter 7)
▶ BS 5266-1 *Emergency lighting* (see also IET publication *Electrician's Guide to Emergency Lighting*)
▶ BS 7346-6 *Components for smoke and heat control: Specifications for cable systems*

Note BS 7346-6 does not currently cover:

a cables of rated voltage not exceeding 600/1000 V with the overall outside diameter less than 20 mm which are not capable of being fitted into the test rig;
b wiring systems where the cables of rated voltage not exceeding 600/1000 V are installed within metallic conduit;
c wiring systems used for fire alarm detection and sounder circuits as described in BS 5839.

Buildings other than dwellings – British Standard 5839-1

7

7.1   Scope of BS 5839-1

(BS 5389-1 Clause 1)

British Standard BS 5839 *Fire detection and fire alarm systems for buildings* is concerned with the design and installation of equipment for fire detection and alarm. It is not concerned with building design. It does not give guidance as to whether or not a fire alarm system should be installed in premises.

The Building Regulations in respect of new buildings and the Regulatory Reform (Fire Safety) Order for existing buildings will normally determine the need for a fire alarm system in a building – see Chapters 1 to 4 of this Guide.

BS 5839-1 provides recommendations for the planning, design, installation, commissioning and maintenance of fire detection fire alarm systems in and around buildings *other than dwellings*. Recommendations for fire detection and alarm systems in dwellings are given in Part 6 of BS 5839 (see Chapter 5 of this Guide).

This chapter is concerned with BS 5839-1 and Volume 2 of Approved Document B.

7.2   Approved Document B

Compliance with BS 5839-1 will generally lead to compliance with both Approved Document B, Volume 1 (dwellinghouses) and Volume 2 (buildings other than dwellinghouses).

Specific references to BS 5839 in Volume 2 of Approved Document B are listed in Table 7.1.

▼ **Table 7.1** References to BS 5839 in Approved Document B, Volume 2

Clause	Requirement
1.4	All new flats should be provided with a fire detection and alarm system in accordance with the relevant recommendations of BS 5839-6:2004 *Code of practice for the design, installation, maintenance and servicing of fire detector and alarm systems in dwellings* to at least Grade D Category LD3 standard.
1.5	More information on power supplies is given in Clause 15 of BS 5839-6.
1.9	**Positioning of smoke and heat alarms** Detailed guidance on the design and installation of fire detection and alarm systems in flats is given in BS 5839-6. However, the following guidance is appropriate to most common situations.
1.18	Note: BS 5839-1 and BS 5939-6 recommend that occupiers should receive the manufacturer's instructions concerning the operation and maintenance of the alarm system.
1.22	**Power supplies** Other effective options exist and are described in BS 5839: Parts 1 and 6. For example, the mains supply may be reduced to extra low voltage in a control unit incorporating a standby trickle-charged battery, before being distributed at that voltage to the alarms.
1.26	Note: General guidance on the standard of automatic fire detection that may need to be provided can be found in Table A1 of BS 5839-1:2002.
1.30	An electrically operated fire alarm system should comply with BS 5839-1:2002 *Fire detection and alarm systems for buildings. Code of practice for system design, installation, commissioning and maintenance.*
1.31	Call points for electrical alarm systems should comply with BS 5839-2:1983 *Specification for manual call points*, or Type A of BS EN 54 *Fire detection and fire alarm systems – Part 11: Manual call points*, and these should be installed in accordance with BS 5839-1. Type B call points should only be used with the approval of the building control body.
1.32	If a voice alarm system is to be installed, it should comply with BS 5839-8 *Code of practice for the design, installation and servicing of voice alarm systems.*
1.33	In certain premises, e.g. large shops and places of assembly, an initial general alarm may be undesirable because of the number of members of the public present. The need for fully trained staff to effect pre-planned procedures for safe evacuation will therefore be essential. Actuation of the fire alarm system will cause staff to be alerted, e.g. by discreet sounders, personal paging systems etc. Provision will normally be made for full evacuation of the premises by sounders or a message broadcast over the public address system. In all other respects, any staff alarm system should comply with BS 5839-1.
1.34	Clause 18 of BS 5839-1:2002 provides details of guidance on the design and installation of fire alarm systems for people with impaired hearing.
1.35	Automatic fire detection and alarms in accordance with BS 5839-1 should be provided in institutional and other residential occupancies.
1.36	Automatic fire detection systems are not normally needed in non-residential occupancies. However, there are often circumstances where a fire detection system in accordance with BS 5839-1 may be needed.
3.47	**Residential care homes** A fire detection and alarm system should be provided in accordance with BS 5839-1:2002.

▼ **Table 7.1**   *continued*

Clause	Requirement
4.12	**Design for vertical escape, provision of refuges** The Emergency Voice Communication system should comply with BS 5839-9:2003 *Fire detection and fire alarm systems for buildings* and consist of Type B outstations which communicate with a master station located in the building control room (where one exists) or adjacent to the fire alarm panel.
4.29e	**Phased evacuation** The building should be fitted with an appropriate fire warning system, conforming to at least the L3 standard given in BS 5839-1 *Fire detection and alarm systems for buildings. Code of practice for system design, installation and servicing.*
4.29f	An internal speech communication system should be provided to permit conversation between a control point at fire service access level, and a fire warden on every storey. In addition, the recommendations relating to phased evacuation provided in BS 5839-1 should be followed where it is deemed appropriate to install a voice alarm. This should be in accordance with BS 5839-8 *Code of practice for the design, installation and servicing of voice alarm systems.*
5.38	**Protected power circuits** In a large or complex building there may be fire protection systems that need to operate for an extended period of time. Further guidance on the selection of cables in such systems is given in BS 5839-1, BS 5266-1 and BS 7346-6.
5.53	Guidance on the provision of smoke detectors in ventilation ductwork is given in BS 5839-1:2002.
7.8	**Raised storage areas** Note 1: Where the lower level is provided with an automatic detection and alarm system meeting the relevant recommendations of BS 5839-1 *Fire detection and alarm systems for buildings. Code of practice for system design, installation and servicing*, then the floor size may be increased to not more than 20 m in either width or length.
9.12	**Extensive cavities** Where the concealed space is an undivided area which exceeds 40 m (this may be in both directions on plan) there is no limit to the size of the cavity if: a. the room and the cavity together are compartmented from the rest of the building; b. an automatic fire detection and alarm system meeting the relevant recommendations of BS 5839-1 *Fire detection and alarm systems for buildings. Code of practice for system design, installation and servicing* is fitted in the building. Detectors are only required in the cavity to satisfy BS 5839-1.
10.13	**Fire dampers** However, in a situation where all occupants of the building can be expected to make an unaided escape and an L1 fire system is installed in accordance with BS 5839-1:2002, exceptions may be made.
18.14	**Mechanical smoke extract**

7

7.3   Categories of system (BS 5389-1 Clause 5)

Because of the great variety of applications of systems, systems are divided into a number of different categories (see Table 7.2). These categories are also used in Approved Document B.

▼ **Table 7.2**   Fire alarm system categories

Category	Description
M	Manual system
L	Automatic fire detection system intended for the protection of life
L1	Automatic protection of life system installed throughout the building
L2	Automatic protection of life system installed only in defined parts of the building including all areas protected by Category L3
L3	Automatic protection of life system installed in i escape routes (including circulation areas corridors and staircases), and ii rooms opening onto escape routes
L4	Automatic protection of life system installed in escape routes (including circulation areas, corridors and staircases)
L5	Automatic protection of life system in which the protected area(s) and/or the location of detectors is designed to satisfy a specific fire safety objective (other than that of a Category L1, L2, L3 or L4 system)
P	Automatic fire detection system intended for the protection of property
P1	Automatic protection of property system installed throughout the building
P2	Automatic protection of property system installed only in defined parts of the building

7.3.1   Category M (manual)

Category M systems are **manual** systems and, therefore, incorporate no automatic fire detectors.

Note: Even in buildings with comprehensive fire detection, manual call points will generally be necessary as people in the vicinity of a fire will normally be aware of the fire and able to raise the alarm by the use of a manual call point before it is detected automatically.

7.3.2   Category L (life)

Category L systems are automatic fire detection systems intended for the protection of **life**. They are further subdivided as follows.

Category L1: systems installed throughout all areas of the building. The objective of a Category L1 system is to offer the earliest possible warning of fire, so as to achieve the longest available time for escape.

Category L2: systems installed only in defined parts of the building. A Category L2 system should include the coverage necessary to satisfy the recommendations of this standard for a Category L3 system. The objective of a Category L2 system is identical to that of a Category L3 system, with the additional objective of affording early warning of fire in specified areas of high fire hazard level and/or high fire risk.

Note: The specified parts of the building may be as few as one or more rooms, or as extensive as, for example, complete floors of the building.

Category L3: systems designed to give a warning of fire at an early enough stage to enable all occupants, other than possibly those in the room of fire origin, to escape safely, before the escape routes are impassable owing to the presence of fire, smoke or toxic gases. In Category L3 systems heat and smoke combustion gas or multi-sensor detectors should be installed in all rooms that open on to the escape routes, except that rooms opening onto corridors of less than 4 m in length need not be protected providing fire-resistant construction including doors separates these corridors from any other section of the escape route. In Category L3 (and L4) systems, smoke detectors or a mixture of smoke and combustion gas detectors should be provided in:

▶ all escape stairways
▶ all corridors
▶ any other areas that form part of the common escape routes.

Note: main and egress stairways normally form part of escape routes and should be treated as escape stairways.

Category L4: systems installed within those parts of the escape routes comprising circulation areas and spaces, such as corridors and staircases. The objective of a Category L4 system is to enhance the safety of occupants by providing warning of smoke within principal escape routes. The installation of detectors in additional areas is not precluded, and the system could then still be regarded as a Category L4 system. In Category L4 systems, smoke detectors or a mixture of smoke and combustion gas detectors should be provided in the following:

▶ all escape stairways
▶ all corridors
▶ any other areas that form part of the common escape routes.

Note: main and egress stairways normally form part of escape routes and should be treated as escape stairways.

Category L5: systems in which the protected area(s) and/or the location of detectors is designed to satisfy a specific fire safety objective (other than that of a Category L1, L2, L3 or L4 system). Such a system could be as simple as one that incorporates a single automatic fire detector in one room (in which outbreak of fire would create undue risk to occupants, either in the room or elsewhere in the building), but the system could comprise comprehensive detection throughout large areas of a building in which, for example, structural fire resistance is less than that normally specified for buildings of that type.

Note: The protection afforded by a Category L5 system may, or may not, incorporate that provided by a Category L2, L3 or L4 system.

7.3.3 Category P (property)

Category P systems are automatic fire detection systems intended for the protection of **property**. They are further subdivided as follows.

Category P1: systems installed throughout all areas of the building. The objective of a Category P1 system is to offer the earliest possible warning of fire so as to minimise the time between ignition and the arrival of fire fighters.

Category P2: systems installed only in defined parts of the building. The objective of a Category P2 system is to provide early warning of fire in areas of high fire hazard level, or areas in which the risk to property or business continuity from fire is high.

7.4 Selection of category

Part 1 of BS 5839 does not specify or recommend which category of systems needs to be installed in any given premises. Any instructions to design and install a fire detection and fire alarm system for a building in accordance with Part 1 of BS 5839 is insufficient for a design to be carried out. The categories given in Part 1 are regarded by the standard as a menu from which purchasers, users, specifiers, enforcing authorities and insurers may select a suitable system for the particular building. However, there is an informative Annex A, reproduced here as Table 7.3.

Note: Annex A of BS 5839-1 is informative – that is, it does not constitute recommendations as such, but provides advice as to what is considered to be reasonable understandings or custom and practice interpretations of the fire safety legislation. Decisions on the appropriate category of fire alarm and detection systems must be confirmed with the authorities enforcing the Building Regulations and the Regulatory Reform Order (RRO).

Note: System category is simply a shorthand means of describing whether the system is manual or automatic, and in the case of automatic systems, the object of the automatic fire detection system. The appropriate extent of automatic fire detection will normally be determined by a fire risk assessment, rather than a rigid application of system category to every building of a specific type or occupancy.

7.5 Departures (variations) from the Standard
(BS 5389-1 Clause 7.1)

7.5.1 Introduction

BS 5839-1 is a Code of Practice and its contents are recommendations and not requirements. The recommendations are based on recognised good practice and are suitable for the majority of applications. However, there may be particular situations where the recommendations are unsuitable and could lead to systems that would be unnecessarily expensive, perhaps incorporating measures that would not be cost-effective or difficult to install. There may also be situations where the recommendations of Part 1 may be ineffective. In these circumstances variations from the recommendations are

▼ **Table 7.3**   Choice of appropriate category of fire detection alarm system (Table A1 of BS 5839-1)

Type of premises	Typical category of system	Comments
Common places of work, such as offices, shops, factories, warehouses and restaurants	M or P2/M or P1/M	A Category M system normally satisfies the requirements of legislation. It is, however, often combined with a Category P system to satisfy the requirements of insurers, as company policy for protection of assets, or to protect against business interruption.
Hotels and hostels	L1 or L2	In bedroom areas, the design requirements are usually based on the recommendations for a Category L3 system. Detectors are, however, typically installed in most other rooms and areas, as a fire in almost any area of the building could pose a threat to sleeping occupants; the system category is, therefore, at least L2. In practice, few, if any, areas are left unprotected and the system category is effectively L1, except that a variation from the recommendations applicable to a Category L1 system may apply to the siting of heat, smoke or carbon monoxide detectors in bedrooms; this often follows the recommendations of **22.3e)** of BS 5839-1 for detectors in a Category L3 system.
Large public houses (no residential accommodation)	M	
Public houses with residential accommodation	L2	
Schools, other than small single-storey schools with less than 160 pupils	M or M/P2 or M/P2/L4 or M/P2/L5	System category is normally based on a fire risk assessment. In many schools, a Category P system is installed to combat the hazard of arson. In schools that are partly occupied at certain times (e.g. during evening classes or community use), a Category L4 or L5 system is sometimes considered appropriate.
Hospitals	L1 (with possible minor variations)	Detailed guidance on areas to be protected and possible variations is given in Technical Memorandum 05-03 in England and Wales and SHTM 82 in Scotland.
Places of assembly (e.g. cinemas, theatres, nightclubs, exhibition halls, museums and galleries, leisure centres and casinos)	M/L1–L4	L1 systems are often provided in large or complex buildings.
Small premises (e.g. accommodating less than 300 persons)		
Other premises		

continues

▼ **Table 7.3** *continued*

Type of premises	Typical category of system	Comments
Transportation terminals	M/L5	
Covered shopping centres	L	The exact design needs to be 'tailor made' and often forms part of a fire engineering solution.
Residential care homes	L1–L3	L1 is regarded as appropriate for large premises.
Prisons	M/L5	
Phased evacuation buildings	L3	
Buildings in which other fire precautions, such as means of escape, depart from recognised guidance	M/L5	Automatic fire detectors are sited in such a way as to compensate for the lower standard in other fire precautions.
Buildings with 'inner rooms', from which escape is possible only by passing through another ('access') room, where there is inadequate vision between the inner room and the access room	M/L5	Smoke detectors are sited in the access room.
Buildings in which automatic fire detection is required to operate other fire protection systems (e.g. magnetic door holders)	M/L5	Care is necessary to ensure that automatic fire detectors are sited such that cool smoke cannot pass below the level of the detectors that cause release of the magnetic door holders, and through the (still open) doors.
Situations in which fire could readily spread from an unoccupied area and prejudice means of escape from occupied areas	M/L4 or M/L5	Custom and practice does not involve siting automatic fire detectors in all unoccupied areas, such as plant rooms and storage areas.
Any building in which automatic fire detection is provided as a requirement of a property insurer or to attract an insurance premium discount	M/P1 or M/P2	

specifically recognised by BS 5839-1 and deemed *fire engineering** and may provide a different solution.

Departures from Part 1 and the system in general, must be approved by the regulatory body.

7.5.2   Arbitrary values

Part 1 recognises that all limits set and all recommendations agreed for inclusion in a standard are somewhat arbitrary. They will be based on good custom and practice, and all the judgement of experts as considered appropriate at the time. However, there are circumstances where departures may be acceptable if they have no significant effect on performance or where rigid adherence would be particularly expensive; or even appropriate if they lead to a better design.

Examples of the arbitrary values include:

a   maximum zone size
b   maximum area of protection disabled in the event of specified fault conditions
c   maximum size of open areas in public buildings above which duplication of sounder circuits is recommended
d   maximum travel distance to the nearest manual call point
e   maximum area of coverage of an automatic fire detector
f   minimum sound pressure levels
g   minimum duration of standby power supplies
h   performance parameters for standard and enhanced fire-resisting cables
i   restrictions on the use of the former type of cable.

7.5.3   Variations
(BS 5839-1 Clause 7.1)

Any variations from the recommendations of BS 5839-1 should be clearly identified in the design documents so that they are obvious to any party from whom approval of the specification of the design is being sought, such as the user, purchaser, enforcing authority.

Installers identifying variations from the Standard whilst carrying out installation or commissioning and not clearly identified in the documented design should ensure these are noted and recorded for subsequent approval by the client.

---

* Fire engineering: the application of scientific and engineering principles, rules [Codes], and expert judgement, based on an understanding of the phenomena and effects of fire and of the reaction and behaviour of people to fire, to protect people, property and the environment from the destructive effects of fire.

7.6   Protection of life   (BS 5839-1 Clause 8.1.2)

BS 5839-1 advises that automatic fire detection might not be necessary if:

a   the objective of the fire alarm system is to protect the occupants of a building no one sleeps in, and

b   fire is likely to be detected by people, before smoke seriously reduces visibility in escape routes,

then a category M system would suffice.

A fire risk assessment would be the usual method of determining this and is generally required by fire legislation.

7.7   Integrity and reliability of circuits   (BS 5839-1 Clause 12)

The Standard recommends that:

a   all critical signal paths

b   final circuits providing mains supply, and

c   extra low voltage supply from any external power source

be protected against mechanical damage and fire.

It is also recommended that critical signal paths be monitored so that faults will be identified quickly. External circuits need to be arranged so that the probability of faults that could prevent the system from giving a fire warning are minimised. Clause 12.2.2 lists the following requirements:

a) A fault on one circuit containing manual call points, fire detectors or fire alarm devices, or a combination of them, should not affect any other circuit.

b) Any fault forming a cross-connection between a detector circuit and a sounder circuit should not affect any circuits other than the two circuits involved.

c) A single short circuit or open circuit fault on an automatic fire detector circuit should neither disable protection within an area of more than 2 000 m^2, nor on more than one floor of the building plus a maximum of five devices (automatic detection, manual call points, sounders or a combination of these) on the floor immediately above and five devices on the floor immediately below that floor.

d) Two simultaneous faults on a manual call point or fire detector circuit should not disable protection within an area greater than 10 000 m^2.

NOTE 1 The areas quoted in c) and d) are relatively arbitrary; in the case of c), the recommendation is based on custom and practice in conventional fire alarm systems over many years. Engineering judgment might determine that, in a given system, minor increases in the areas quoted do not significantly impede the system integrity, but this should then be regarded as variation from the recommendations of the standard.

e) For software controlled Control and Indicating equipment that has more than 512 detectors and/or manual call points connected, reference should be made to the manufacturer's instructions regarding the means by which compliance with BS EN 54-2 should be achieved.

f) Where detectors are designed to be removable from their bases for the purpose of routine maintenance of the system:

1) removal of any detector(s) from the circuit should not affect the operation of any manual call point, regardless of whether locking devices are used to secure the detectors within their bases;

NOTE 2 This recommendation applies even if all fire detectors in the system are removed from circuit.

2) during the design stage, consideration should be given to the possibility of malicious removal of detectors. If malicious removal is considered likely, detectors of a type that can be removed only by the use of special tool or special technique should be used.

NOTE 3 A special tool in this context is a tool not likely to be carried by a member of the general public. Slot-headed screws would not be acceptable, since various articles can be used as screwdrivers.

g) Any facility specifically provided for deliberate disablement of manual call point or detector circuits should be such that it is possible to disable protection throughout one zone of the system without disabling protection in other zones. Use of such facilities should not prevent evacuation of the building by use of an evacuate control on, or close to, the control and indicating equipment.

h) Removal of any manual call point or detector from its circuit should not affect the ability of any fire alarm device to respond to an alarm signal other than in the case of devices that incorporate a fire alarm sounder and detector in the same unit.

i) Fire alarm devices should be capable of being disconnected from their circuits only by the use of a special tool.

j) In the event of a single open circuit or short circuit fault on any circuit that serves fire alarm sounders, at least one single fire alarm sounder, located in the vicinity of the control and indicating equipment, should still sound correctly if a fire alarm condition occurs anywhere within the building. This fire alarm sounder should have an identical sound to the general fire alarm sounders in the building.

k) In buildings designed to accommodate the general public in large numbers (e.g. transport terminals, shopping centres, places of public entertainment, department stores and leisure centres), at least two sounder circuits should be provided in every uncompartmented public space if the space is either:

i)   greater than 4,000 m^2 in area; or
ii) designed to accommodate more than 500 members of the public.
This should be achieved either: …

l) Where two (or more) sounder circuits are necessary in order to satisfy the recommendations of j) or k), the circuits should not be contained within a common cable sheath. For example, two circuits intended to satisfy these recommendations should not be served by a common four core cable, as this would not adequately protect against the simultaneous loss of both circuits.

m) If the control and indicating equipment is supplied with power from power supply equipment contained in a separate enclosure, the cables between the equipment should be duplicated such that a single open or short circuit in the connections does not completely remove power from the control and indicating equipment. The duplicate cables should be separated by at least 300 mm where practical.

Where a power supply unit or a standby battery(ies) are housed in a separate enclosure from the control and indicating equipment, any cable between that enclosure and the CIE should be suitably protected against overcurrent in accordance with BS 7671.

7.8 Evacuation strategy (BS 5839-1 Clauses 14.1, 19.1)

The strategy for the evacuation of the building must be decided before design of the fire detection and alarm system is started. The strategy in many buildings will be simple. On the operation of a manual call point or automatic detection of fire, the fire alarms will 'sound' throughout the building and all the occupants expected to evacuate the building via the emergency exits. Designers must know the routes to be taken to ensure that they are appropriately indicated. The users (employer) will also need to understand the strategy so that they can ensure that the escape routes are kept clear.

In larger or complex buildings it may be necessary to stage the evacuation and the initiation of sounders or other evacuation signals such as voice indications will need to take account of this.

When deciding upon the alarm zones the designer will need to take account of the fire-resisting properties of the building. Zones must take account of likely fire spread. The evacuation strategy should be as simple as is possible to ensure safe evacuation in the stressful circumstances of the fire, when in the practical circumstances certain escape routes may be blocked and voice alarm advice unclear etc.

7.9 Power supplies

7.9.1 Introduction (BS 5839-1 Clause 25.2)

Generally, the supply to the fire system and the circuits of the fire system are to be protected from faults, overload, switching operations, and general use of the remainder of the electrical installation. However, it is presumed that there will be failures, and as a consequence standby supplies are required. Unless there are particular circumstances, the standby supplies should be able to run the fire system for 24 hours plus 30 minutes operation of the alarm devices – see section 7.9.4 below. Where there is to be staged evacuation or other special circumstances, the standby time, and more particularly the operation time, may need to be increased so that the escape systems remain operational for the required evacuation time.

Control and indicating equipment must give an indication of loss of supply to the fire system as failure of supply to the fire system may occur without complete loss of supply in the building.

Note: BS 7671 (see Chapter 8) allows protection against overload to be omitted (regulation 433.3.3 and 560.7.3) to provide for a more reliable supply. This allows overcurrent devices (fuses and circuit-breakers to be 'oversized' so that unwanted operation is avoided. As the electrical load of the fire system will not change (without modification), this is within the general rules. Fault protection and shock protection must be provided.

7.9.2   Segregation of fire systems
(BS 5839-1 Clause 25.1, BS 7671 regulation 560.7)

The design of the electrical installation in the building must minimise the risk of faults affecting fire alarm and escape systems. Safety circuits including supplies to fire systems are to be independent of other circuits. This is a recommendation of BS 5839-1 and a requirement of regulation 560.7 of BS 7671.

It is important that supplies to the fire alarm installation can be maintained when the rest of the electrical installation is isolated for maintenance purposes or perhaps for economy in the consumption of electricity (see Figure 7.1).

▼ **Figure 7.1**   Separation of fire system

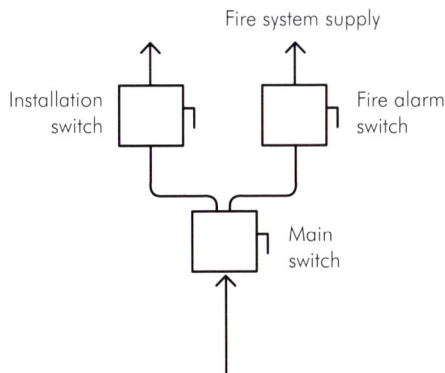

BS 5839 requires that the mains supply be backed up by standby supply arrangements that are able to support the fire alarm system for a minimum of 24 hours, see section 7.9.4 below.

The installation of a standby supply is also a requirement of the Health and Safety (Safety Signs and Signals) Regulations 1996 – see paragraph 8 of Schedule 1 (reproduced below).

SCHEDULE 1 Regulation 4(4) and (5)

PART I
MINIMUM REQUIREMENTS CONCERNING SAFETY SIGNS AND SIGNALS AT WORK

8. Signs requiring some form of power must be provided with a guaranteed emergency supply in the event of a power cut, unless the hazard has thereby been eliminated.

9. The triggering of an illuminated sign and/or acoustic signal indicates when the required action should start; the sign or signal must be activated for as long as the action requires. Illuminated signs and acoustic signals must be reactivated immediately after use.

10. Illuminated signs and acoustic signals must be checked to ensure that they function correctly and that they are effective before they are put into service and subsequently at sufficiently frequent intervals.

Note: Signs are safety signs and include fire safety signs (these may be an illuminated sign or an acoustic signal).

Unless the presumption is unreasonable in the particular circumstances, it is to be assumed that failures exceeding 24 hours are unlikely and that the duration of standby batteries need therefore not be capable of supplying to the system for longer than 24 hours. There may be circumstances where this is an unreasonable assumption and in this case adequate precautions will need to be taken. This may be the situation in remote locations where it is known from experience that outages exceeding 24 hours are to be expected.

If the premises are provided with an automatically started standby generator, the capacity of the standby batteries provided for the fire system in a Category M (manual) or L (protection of life) system may be reduced (if the fire alarm systems are indeed backed up by the standby system). The presumption is that people will be present to detect and take action on the failure of the standby generator. This would not be appropriate for category P premises that may be unoccupied.

7.9.3 Mains power supplies (BS 5839-1 Clause 25.2)

BS 5839-1 includes requirements for low-voltage mains power supplies to the fire system to improve reliability, avoid unintended switching off and facilitate maintenance as follows.

1 The supply to the fire system is taken from the load side of the main switch (see Figure 7.1). Where the user is likely to want to regularly switch off the supply to the building, for example during closed hours, the supply arrangement should be such that this can be done without switching off the supply to the alarm system.

2 The connection of the fire alarm system should be made close to the main intake. Where practicable, there should be a dedicated fire alarm supply circuit.

3 A local isolator, switching both the phase and neutral should be installed adjacent to the fire alarm panel.

4 The number of switches/isolators should be kept to a minimum.

5 Every switch and protective device that can switch off the supply to the fire alarm system (other than the main switch) should be labelled as follows (see Figure 7.2):

a in the case of a protective device that serves only the fire alarm circuit, but incorporates no switch

> FIRE ALARM

b in the case of a switch (whether incorporating a protective device or not) that serves only the fire alarm circuit

> FIRE ALARM. DO NOT SWITCH OFF

c in the case of any switch that disconnects the mains supply to both the fire alarm system and to other circuits (Note: this should not normally be the arrangement).

> WARNING. THIS SWITCH ALSO CONTROLS THE SUPPLY
> TO THE FIRE ALARM SYSTEM

Labels should be clear and in durable fade-resistant material.

▼ **Figure 7.2**   Labelling of electrical supply

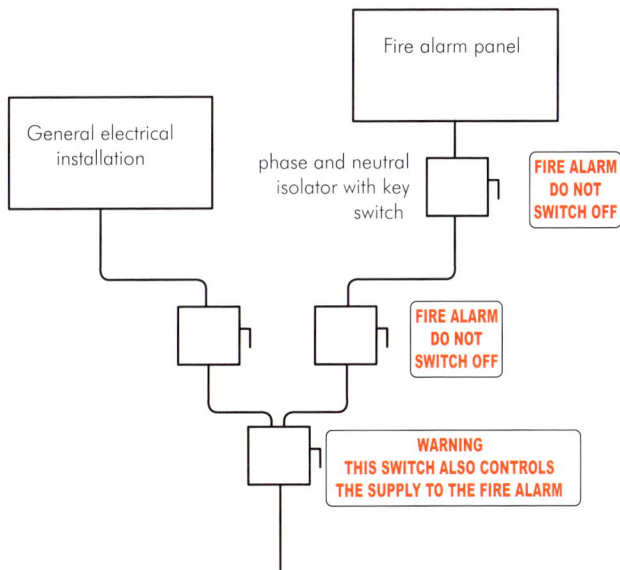

6 Every switch and protective device that could switch off the supply to the fire alarm system should be:

a situated in a location accessible only to authorised persons, or

b protected against unauthorised operation by a key switch or similar.

7   The circuit supplying the fire alarm system should not be protected by a residual current device (including an RCBO) unless this cannot be avoided to comply with the requirements of BS 7671 such as in a TT installation. Where a residual current device is necessary, it must be segregated from the remainder of the installation such that a fault in the general installation will not result in disconnection of the supply to the fire system (see Figure 7.3).

8   The mains power supply must be capable of supplying the maximum load of the installation including the fire system without support from the standby supply.

▼ **Figure 7.3**   TT installation (with RCBOs)

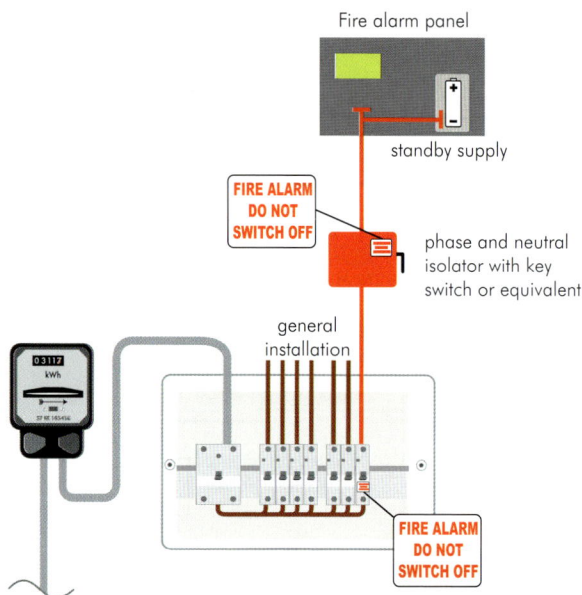

7.9.4   **Fire alarm system power supply units**   (BS 5839-1 Clause 25.3)

The following recommendations apply to every power supply unit that forms part of the fire alarm system.

1   Switching from the normal supply to the standby supply, and vice versa, should not cause any interruption to the operation of the system or result in a false alarm.

2   The fire system should be segregated from the general installation so that faults on one will not affect the other. The operation of a single protective device should not result in failure of both the normal and the standby supply.

3   The presence of the normal or the standby supply should be shown by a green indicator, sited in a readily visible position (e.g. at the the main indicating equipment).

4   Normal and standby supplies should each be independently capable of supplying the maximum alarm load of the system.

7.9.5   Recommendations for standby supplies   (BS 5839-1 Clause 25.4)

The summarised requirements of Clause 25.4 are listed below. For full details consult the standard.

a   The standby supply should comprise rechargeable battery with an automatic charger.
b   The battery should have a life of at least four years in use.
c   Batteries should be visibly labelled with date of installation.
d   The charging rate of the battery should recharge the battery in 24 hours.

Note: This is the requirement within BS EN 54-4, but alterations to the system including upgrading the battery may require a check of the charger.

e   The capacity of all standby batteries should be calculated as per Annex D of BS 5839-1 so that:

i   for a Category M or Category L system, the capacity should be sufficient to maintain the system in operation for at least 24 h, after which sufficient capacity should remain to provide an 'Evacuate' signal in all alarm zones for at least 30 min.

Note: For a Category M or Category L system in a building with an automatically started standby generator that serves the fire alarm system, the capacity should be sufficient to maintain the system in operation for at least 6 h, after which sufficient capacity should remain to provide an 'Evacuate' signal in all alarm zones for at least 30 min. If there are any parts of the system not supplied by the standby generator supply, 24 h capacity is required.

ii   For a Category P system, the capacity should be sufficient to maintain the system in operation for at least 24 h longer than the maximum period for which the premises are likely to be unoccupied or for 72 h in total, whichever is less, after which sufficient capacity should remain to operate all fire alarm devices for at least 30 min.

7.9.6   Cables, wiring and other interconnections   (BS 5839-1 Clause 26)

Selection of cables

The Standard recognises the two cable types as follows:

a   standard fire-resisting cables
b   enhanced fire-resisting cables.

Standard fire-resisting cables are to be used for all parts of the critical signal paths and for the final circuits to the fire system.

Note: Definition of *critical signal path*: all components and interconnections between every fire alarm initiation point (manual call point or automatic fire detector) and the input terminals on or within each fire alarm device.

Enhanced fire-resisting cables are recommended for systems in which cables might need to operate satisfactorily in:

a   Unsprinklered buildings in which the fire strategy involves evacuation of occupants in four or more phases.

b Unsprinklered buildings greater than 30 m in height.

c Unsprinklered premises and sites in which a fire in one area could affect cables of critical signal paths associated with areas remote from the fire, in which it is envisaged people will remain in occupation during the course of the fire. Examples may be large hospitals with central control equipment and progressive horizontal evacuation arrangements, and certain large industrial sites.

d In any other buildings in which the designer, specifier or regulatory authority, on the basis of a fire risk assessment that takes fire engineering considerations into account, considers that the use of enhanced fire-resisting cables is necessary.

Standard fire-resisting cables (used for critical signal paths) (BS 5839-1 Clause 26.2)

Standard fire-resisting cables should:

1 meet the PH 30 (or PH 100) classification when tested in accordance with BS EN 50200 (and BS 8434-1), see note below, and

2 be selected from one of the following:

    a mineral insulated copper sheathed cables, with an overall polymeric covering, conforming to BS EN 60702-1, with terminations conforming to BS EN 60702-2

    b fire-resistant cables that conform to BS 7629

    c armoured fire-resistant cables that conform to BS 7846

    d cables rated at 300/500 V (or greater) that provide the same degree of safety to that afforded by compliance with BS 7629.

Note: Standard fire-resisting cables are required to meet the PH 30 classification and maintain circuit integrity when tested in accordance with BS EN 50200 and tests of simultaneously exposing a sample of the cable to flame at a temperature of 830 °C and mechanical shock for 15 min, then simultaneously exposing it to water spray and mechanical shock for a further 15 min.

In practice the installer needs confirmation from the cable supplier that cables selected meet the requirements of BS 5839-1 for standard fire-resisting cables.

Enhanced fire-resisting cables

Enhanced fire-resisting cables are required to meet the PH 120 classification when tested in accordance with BS EN 50200 and maintain circuit integrity when subjecting a sample of the cable to flame at a temperature of 930 °C and mechanical shock for a period of 60 min, then simultaneously exposing it to water spray and mechanical shock for a further 60 min.

As the test requirements for cables of BS 5839-1 do not align with the requirements of the cable standards, specifiers and installers must confirm with the cable supplier that the requirements are met.

Table 7.4 shows typical examples of a manufacturer's statements on compliance.

▼ **Table 7.4**   A manufacturer's statements on compliance

FP200 Gold	BASEC and LPCB approved to the BS5839-1:2002 requirements for 'standard' cables
FP PLUS™	BASEC and LPCB approved to the BS 5839-1:2002 requirements for 'enhanced' cables
FP PLUSFLEX	BASEC and LPCB approved to the BS 5839-1:2002 requirements for 'enhanced' cables
FP400	BASEC approved to BS 7846 LPCB, approved to BS 6387 category CWZ
FP600	Tested to BS 7346-6:2005 *Components for smoke and heat control systems – Part 6 Cable systems* 120-minute rating

Cable colours

All fire system cables should be of a common colour, different to that used for the cables of the general electrical installation, and the preferred colour is red.

Cable installation

The standard makes the following recommendations for the installation of cables:

1 Cables should be installed without external joints, wherever practicable.
2 Mineral insulated copper sheathed cables and steel wire armoured cables may be used throughout without additional mechanical protection (note or protection by RCDs).
3 Other cables should be given mechanical protection in any area, where physical damage or rodent attack is a risk. Particular attention is to be given to cables installed less than 2 m above the floor level.
Note: Mechanical protection includes use of: cable tray, conduit, ducting, trunking and burying in the structure of the building. In many cases the fire-resisting cable is contained within a trunking system. In such cases the cable must be securely fixed to the building structure through the trunking itself.
4 Cable fixings and supports must be such as to match the fire, water and shock withstands specified for the cables.
5 Penetrations of walls, ceilings, floors must be sealed to maintain fire and smoke integrity.
6 Fire system cables shall be segregated from other cables.*
7 ELV cables shall be segregated from LV cables.

Note: Clause 37 of BS 5839-1 has good practice recommendations that the installer will need to follow.

---

* The use of separate fire-resistant cables and separate conduits and trunking (or metal partitions) is appropriate.

7.10   RCDs

BS 7671 requires cables installed in walls, partitions, floors or ceilings at a depth of less than 50 mm to:

a incorporate a metal covering, or
b be enclosed in earthed steel conduit or trunking, or
c be otherwise mechanically protected against penetration by nails or screws or the like, or
d in the case of walls and partitions, be installed in defined zones (see Figure 7.4).

▼ **Figure 7.4**   Permitted cable routes

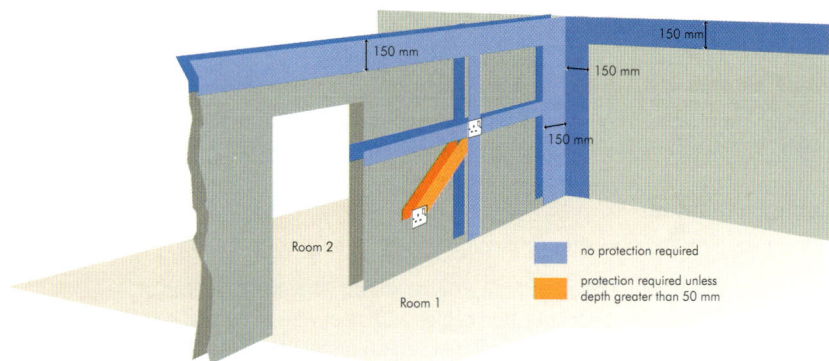

150 mm
150 mm
150 mm
150 mm
150 mm

Room 2
Room 1

no protection required

protection required unless depth greater than 50 mm

For metal-framed walls and partitions, protection as a to c above is required at any depth and in the zones.

Additionally, where the installation is not intended to be under the supervision of a skilled or instructed person, cables without the mechanical protection of **a** to **c** above should be protected by a 30 mA RCD or RCBO.

Most commercial and industrial installations will be intended to be under the supervision of a skilled or instructed person. The Electricity at Work Regulations will require the employer to provide such supervision to prevent for example the use of unauthorised equipment and the drilling of walls etc. by untrained persons. However, designers can avoid this decision by:

a using cables incorporating a metal covering and with fire resistance, or
b as BS 5839-1 sensibly suggests, by fixing the cable to the surface.

The following cable standards are quoted in Clause 522.6.5 and 522.6.6 of BS 7671:

a Mineral insulated copper sheathed cables, with an overall polymeric covering, conforming to BS EN 60702-1, with terminations conforming to BS EN 60702-2.
b Armoured fire-resistant cables that conform to BS 7846.

7.11 Sound levels (BS 5839-1 Clause 16)

For manual and automatic alarm systems (Categories M and L) the Standard sets minimum sound pressure levels as follows:

▶ All accessible areas of the building – not less than 65 dB(A)
▶ Stairways, small enclosures (no more than 60 m²), areas of limited extent – not less than 60 dB(A)

Note: Sound levels in locations used to contact the fire and rescue sevices should not be so loud as to prevent conversation.

7.12 Radio linked systems (BS 5839-1 Clause 27)

The Standard recognises that radio linked systems cannot comply with all the requirements for hard-wired systems and that there are advantages and disadvantages in both systems. It is likely that the demand for radio-linked systems will grow, particularly because of the relative ease of installation in existing buildings.

Systems cannot be completely designed from drawings; it is necessary to test on site. Booster aerials may be required on completion.

Radio links can be used for almost all elements of the system including detectors, manual call points, and sounders. Their advantages and disadvantages are summarised in Table 7.5.

▼ **Table 7.5**   Advantages and disadvantages of radio linked systems

Advantages	Comments
Reduced hard wiring	Particularly easy to install in existing buildings
Flexible	Easy to change locations and add elements (sounders, call points etc.)
Disadvantages	**Comments**
Interference	
Variable signal strength	Need to test on site before and after installation
Low-frequency electromagnetic radiation	Some customers may have concerns
Reliance on batteries	Cost of periodic battery replacement

Radio-linked systems must be designed and equipment specified to comply with Clause 27.2 of BS 5839-1 and BS EN 54-25 *Fire detection and fire alarm systems. Components using radio links*. This includes a requirement to comply (until a standard IEC, CENELEC or BS is available) with Loss Prevention Certificate Board test standard LPS 1257, as amended by Clause 27.2a) of BS 5389-1.

The exceptions/amendments of Clause 27.2b) of BS 5839-1 are as follows:

a All radio-linked components should be supplied from at least two independent power supplies. These can be either:

   i the normal mains supply plus a reserve battery (primary or continuously charged secondary); or

   ii a primary battery plus a second primary battery; or

   iii a primary battery plus a secondary battery.

b Components, other than control and indicating equipment, may utilise batteries to provide the normal power supply.

c Power supplies incorporating one or more primary batteries should give at least 30 days warning of impending failure of each battery. This should be indicated as a low battery warning condition at the control and indicating equipment.

d At the point at which the power supply(ies) to any radio-linked component can maintain the component in normal operation for no more than 7 days, and, in addition, in the case of the fire alarm devices, 30 min in the alarm condition, a fault warning should be given at the control and indicating equipment.

e Power supplies should have a minimum normal operational life of 3 years over the temperature range of +15 °C to +35 °C before the low power condition is signalled.

f Any fault giving rise to loss of communication with a radio-linked component should be indicated at the control and indicating equipment within 2 h of occurrence of the fault.

7.13 Inspection, testing and certification (Clause 38 and Annex G)

7.13.1 Testing

Clause 38 of BS 5839 generally requires inspection and testing in accordance with BS 7671.

BS 7671 requires inspection and testing and the issue of certificates and inspection and test schedules. These must be provided in accordance with Part 6 and Appendix 6 to the person ordering the work. These will be in addition to the certificates of BS 5839.

7.13.2 Certification

The model certificates in this section are taken from Annex G of BS 5839-1.

G.1 Design certificate

Certificate of design for the fire alarm system at:

Address: ....

....

I/we being the person(s) responsible (as indicated by my/our signatures below) for the design of the fire alarm system, particulars of which are set out below, CERTIFY that the said design for which I/we have been responsible complies to the best of my/our knowledge and belief with the recommendations of Section 2 of BS 5839-1: 2001 for the system category described below, except for the variations, if any, stated in this certificate.

Name (in block letters): .... Position: ....

Signature: .... Date: ....

For and on behalf of: ....

Address: ....

....

.... Postcode: ....

The extent of liability of the signatory is limited to the system described below.

System category (see BS 5839-1 Clause 5): ....

Variations from the recommendations of Section 2 of BS 5839-1 (see BS 5839-1 Clause 7):

....

....

....

....

Extent of system covered by this certificate: ....

....

Brief description of areas protected (not applicable for Category M, L1 or P1 systems):

....

....

....

....

Measures incorporated to limit false alarms. Account has been taken of the guidance contained in section 3 of BS 5839-1: 2002 and, more specifically (tick as appropriate):

☐ The system is manual. Type and siting of manual call points takes account of the guidance contained in Section 3 of BS 5839-1.

☐ The system incorporates automatic fire detectors, and account has been taken of reasonably foreseeable causes of unwanted alarms, particularly in the selection and siting of detectors.

☐ An appropriate analogue system has been specified.

☐ An appropriate multi-sensor system has been specified.

☐ A time-related system has been specified. Details: ....

....

☐ Fire signals from automatic fire detectors result initially in a staff alarm, which delays a general alarm/transmission of signals to an alarm receiving centre (delete as applicable) for .... min.

☐ Appropriate guidance has been provided for the user to enable limitation of false alarms.

☐ Other measures as follows: ....

....

Installation and commissioning

It is strongly recommended that installation and commissioning be undertaken in accordance with the recommendations of Sections 4 and 5 of BS 5839-1:2002 respectively.

Soak test

☐ In accordance with the recommendations of Clause 35.2.6 of BS 5839-1:2002, it is recommended that, following commissioning, a soak period of .... should follow. (Enter a period of not less than one week.)

☐ As the system incorporates no more than 50 automatic fire detectors, no soak test is necessary to satisfy the recommendations of BS 5839-1: 2002.

Verification

Verification that the system complies with BS 5839-1: 2001 should be carried out, on completion, in accordance with Clause 43 of BS 5839-1:

Yes ☐   No ☐   To be decided by the ☐
                              purchaser or user

Maintenance

It is strongly recommended that, after completion, the system is maintained in accordance with Section 6 of BS 5839-1: 2002.

User responsibilities

The user should appoint a responsible person to supervise all matters pertaining to the fire alarm system in accordance with the recommendations of Section 7 of BS 5839-1: 2002.

G.2 Installation certificate

Certificate of installation for the fire alarm system at:

Address: ....
....

I/we being the person(s) responsible (as indicated by my/our signatures below) for the installation of the fire alarm system, particulars of which are set out below, CERTIFY that the said installation work for which I/we have been responsible complies to the best of my/our knowledge and belief with the specification described below and with the recommendations of Section 4 of BS 5839-1: 2002, except for the variations, if any, stated in this certificate.

Name (in block letters): .... Position: ....

Signature: .... Date: ....

For and on behalf of: ....

Address: ....

....

.... Postcode: ....

The extent of liability of the signatory is limited to the system described below.

Extent of installation work covered by this certificate:
....
....
....
....

Specification against which system was installed:
....
....
....
....

Variations from the specification and/or Section 4 of BS 5839-1 (see BS 5839-1 Clause 7):
....
....
....

Wiring has been tested in accordance with the recommendations of Clause 38 of BS 5839-1: 2002. Test results have been recorded and provided to: ....

Unless supplied by others, the 'as fitted' drawings have been supplied to the person responsible for commissioning the system (see **36.2** m of BS 5839-1: 2002).
....

G.3 Commissioning certificate

Certificate of commissioning for the fire alarm system at:

Address: ....

....

I/we being the person(s) responsible (as indicated by my/our signatures below) for the commissioning of the fire alarm system, particulars of which are set out below, CERTIFY that the said work for which I/we have been responsible complies to the best of my/our knowledge and belief with the recommendations of clause 39 of BS 5839-1:2002, except for the variations, if any, stated in this certificate.

Name (in block letters): .... Position: ....

Signature: .... Date: ....

For and on behalf of: ....

Address: ....

....

.... Postcode: ....

The extent of liability of the signatory is limited to the system described below.

Extent of system covered by this certificate:

....

....

....

....

Variations from the recommendations of Clause 39 of BS 5839-1: 2002 (see BS 5839-1: 2002 Clause 7):

....

....

....

☐ All equipment operates correctly.

☐ Installation work is, as far as can reasonably be ascertained, of an acceptable standard.

☐ The entire system has been inspected and tested in accordance with the recommendations of 39.2 c) of BS 5839-1: 2002.

☐ The system performs as required by the specification prepared by: .... a copy of which I/we have been given.

☐ Taking into account the guidance contained in section 3 of BS 5839-1: 2002, I/we have not identified any obvious potential for an unacceptable rate of false alarms.

☐ The documentation described in Clause 40 of BS 5839-1:2002 has been provided to the user.

The following work should be completed before/after (delete as applicable) the system becomes operational:

....

....

....

....

The following potential causes of false alarms should be considered at the time of the next service visit:

....

....

....

Before the system becomes operational, it should be soak tested in accordance with the recommendations of 35.2.6 of BS 5839-1: 2002 for a period of: ....
(Enter a period of either one week, such period as required by the specification, or such period as recommended by the signatory to this certificate, whichever is the greatest, or delete if not applicable.)

G.4 Acceptance certificate

Certificate of acceptance for the fire alarm system at:

Address: ....
....

I/we being the person(s) responsible (as indicated by my/our signatures below) for the acceptance of the fire alarm system, particulars of which are set out below, ACCEPT the system on behalf of:

Name (in block letters): .... Position: ....

Signature: .... Date: ....

For and on behalf of: ....

Address: ....
....
.... Postcode: ....

The extent of liability of the signatory is limited to the system described below.

Extent of system covered by this certificate:
....
....

☐ All installation work appears to be satisfactory.
☐ The system is capable of giving a fire alarm signal.
☐ The facility for remote transmission of alarms to an alarm receiving centre operates correctly. (Delete if not applicable.)

The following documents have been provided to the purchaser or user:

☐ 'As fitted' drawings.
☐ Operating and maintenance instructions.
☐ Certificates of design, installation and commissioning.
☐ A log book.
☐ Sufficient representatives of the user have been properly instructed in the use of the system, including, at least, all means of triggering fire signals, silencing and resetting the system and avoidance of false alarms.
☐ All relevant tests, defined in the purchasing specification, have been witnessed. (Delete if not applicable.)

The following work is required before the system can be accepted:
....
....
....
....

G.5 Verification certificate (optional)

Certificate of verification for the fire alarm system at:

Address: ....

....

I/we being the person(s) responsible (as indicated by my/our signatures below) for the verification of the fire alarm system, particulars of which are set out below, CERTIFY that the verification work for which I/we have been responsible complies to the best of my/our knowledge and belief with the recommendations of Clause 43 of BS 5839-1: 2002.

Name (in block letters): .... Position: ....

Signature: .... Date: ....

For and on behalf of: ....

Address: ....

....

.... Postcode: ....

The extent of liability of the signatory is limited to the system described below.

Extent of system covered by this certificate:

....

....

Scope and extent of the verification work:

....

....

....

....

☐ In my/our opinion, as far as can reasonably be ascertained from the scope of work described above, the system complies with, and has been commissioned in accordance with, the recommendations of BS 5839-1: 2002, other than in respect of variations already identified in the certificates of design, installation or commissioning.

☐ In my/our opinion, there is no obvious potential for an unacceptable rate of false alarms.

The following non-compliances with the recommendations of BS 5839-1: 2002, have been identified (other than those recorded as variations in the certificates of design, installation or commissioning):

....

....

....

G.6 Inspection and Servicing certificate

Certificate of servicing for the fire alarm system at:

Address: ....
....

I/we being the person(s) responsible (as indicated by my/our signatures below) for the servicing of the fire alarm system, particulars of which are set out below, CERTIFY that the said work for which I/we have been responsible complies to the best of my/our knowledge and belief with the recommendations of Clause 45 of BS 5839-1: 2002 for quarterly inspection of vented batteries/inspection and test/periodic inspection and test over a 12 month period (delete as applicable), except for the variations, if any, stated in this certificate.

Name (in block letters): .... Position: ....

Signature: .... Date: ....

For and on behalf of: ....

Address: ....
....
.... Postcode: ....

The extent of liability of the signatory is limited to the system described below.

Extent of system covered by this certificate:
....
....

Variations from the recommendations of Clause 45 of BS 5839-1: 2002 for periodic or annual inspection and test (as applicable):
....
....
....
....

☐ Relevant details of the work carried out and faults identified have been entered in the system log book.

During the past 12 months .... false alarms have occurred.

The above number of false alarms equates to ....false alarms per 100 automatic fire detectors per annum (for Category M systems enter 'Not applicable').

The following work/action is considered necessary:
....
....
....
....
....
....

G.7 Modification certificate

Certificate of modification for the fire alarm system at:

Address: ....

....

I/we being the person(s) responsible (as indicated by my/our signatures below) for the modification of the fire alarm system, particulars of which are set out below, CERTIFY that the said modification work for which I/we have been responsible has to the best of my/our knowledge and belief been carried out in accordance with the recommendations of **46.4**, of BS 5839-1: 2002, except for the variations, if any, stated in this certificate.

Name (in block letters): .... Position: ....

Signature: .... Date: ....

For and on behalf of: ....

Address: ....

....

.... Postcode: ....

The extent of liability of the signatory is limited to the system described below.

Extent of system covered by this certificate:

....

....

Variations from the recommendations of 46.4 of BS 5839-1: 2002:

....

....

....

....

☐ Following the modifications, the system has been tested in accordance with the recommendations of 46.4.2 of BS 5839-1: 2002

☐ Following the modifications, 'as fitted' drawings and other system records have been updated as appropriate.

I/we the undersigned confirm that the modifications have introduced no additional variations from the recommendations of BS 5839-1:2002, other than those recorded below:

Signed: ....

Capacity: ....

(e.g. maintenance organisation, system designer, consultant or user representative)

7.14   Maintenance (Section 6 of BS 5839-1)

7.14.1 Introduction

BS 5839-1 recommends weekly* testing and inspection by the user and professional attention 6 monthly and annually.

Care should be taken to ensure that any person undertaking these tasks is competent to do so safely and has the relevant technical knowledge and training.

Note: Some fire detection and fire alarm systems and components include features that permit functions to be automatically monitored, and faults or warnings to be annunciated, or otherwise made available to authorised persons. In such cases, the recommendations for routine testing may be modified to omit testing which is declared to be unnecessary by the equipment supplier, provided it can be demonstrated that the automatic monitoring achieves the same objective as the appropriate test recommended.

7.14.2 Weekly user testing (BS 5839-1 Clause 44)

Note: When testing the fire detection system, there may be a need to isolate ancillary outputs.

a   A manual call point should be operated during normal working hours. It should be confirmed that:
  ▶ the control equipment is capable of processing a fire alarm signal and providing an output to fire alarm sounders; and
  ▶ the fire alarm signal is correctly received at any alarm receiving centre to which fire alarm signals are transmitted.

Notes:

i   It is not necessary to confirm that all fire alarm sounder circuits operate correctly at the time of this test.

ii   It is essential that any alarm receiving centre is contacted immediately before, and immediately after, the weekly test to ensure that unwanted alarms are avoided and that fire alarm signals are correctly received at the alarm receiving centre.

iii   The user needs to take account of the manufacturer's recommendations, particularly when battery-powered devices are being tested, e.g. within wireless fire alarm systems.

iv   The test should be carried out at approximately the same time each week; instructions to occupants should then be that they should report any instance of poor audibility of the fire alarm signal. In systems with staged alarms incorporating an 'Alert' and an 'Evacuate' signal, the two signals should be operated, where practicable, sequentially in the order they would occur at the time of a fire (i.e. 'Alert' and then 'Evacuate').

v   In premises in which some employees only work during hours other than that at which the fire alarm system is normally tested, an additional test(s) should be

---

* Monthly maintenance is additionally required in the unusual use of open-cell batteries or the use of a standby generator.

carried out at least once a month to ensure familiarity of these employees with the fire alarm signal(s).

vi A different manual call point should be used at the time of every weekly test, so that all manual call points in the building are tested in rotation over a prolonged period. There is no maximum limit for this period (e.g. in a system with 150 manual call points, the user will test each manual call point every 150 weeks). The result of the weekly test and the identity of the manual call point used should be recorded in the system log book.

vii The duration for which any fire alarm signal is given (other than solely at control and indicating equipment) at the time of the weekly test by the user should not normally exceed 1 min, so that, in the event of a fire at the time of the weekly test, occupants will be warned by the prolonged operation of the fire alarm devices.

b Voice alarm systems should be tested weekly in accordance with the recommendations of BS 5839-8.

7.14.3 Monthly user attention

a Any automatically started emergency supply should be started up by simulation of failure of the normal power supply and operated on-load for at least 1 h. At the end of the test, the fuel tanks should be left filled, and the oil and coolant levels should be checked and topped up as necessary.

Note: The test should be carried out in accordance with the instructions of the manufacturer, including instructions on the load that should be operated.

b Any vented batteries used as a standby power supply should be visually inspected to check that they are in good condition. Any defect, including low electrolyte level, should be rectified.

7.14.4 Quarterly professional inspection of vented batteries   (BS 5839-1 Clause 45.2)

All vented batteries and their connections should be examined by a person competent in battery installation and maintenance technology. Electrolyte levels should be checked and topped up as necessary.

Note: In many large premises and sites, in-house maintenance personnel may be competent to carry out this task.

7.14.5 Six-monthly professional periodic inspection and test   (BS 5839-1 Clause 45.3)

Note: The period between successive inspection and servicing visits should be based upon a risk assessment, taking into account the type of system installed, the environment in which it operates and other factors that may affect the long-term operation of the system. The recommended period between successive inspection and servicing visits should not exceed 6 months. If a risk assessment shows a need for more frequent inspection and servicing visits, then all interested parties should agree the appropriate inspection and servicing schedule. If this recommendation is not implemented, it should be considered that the system is no longer compliant with this part of BS 5839-1.

a   The system log book is examined to confirm that any faults recorded have been rectified.

b   A visual inspection is made to check whether structural or occupancy changes have affected the compliance of the system with the recommendations of the Standard for the siting of manual call points, automatic fire detectors and fire alarm devices. Particular care should be taken to determine if:

  i   all manual call points remain unobstructed and conspicuous;

  ii   any new exits have been created without the provision of an adjacent manual call point;

  iii   any new or relocated partitions have been erected within 500 mm horizontally of any automatic fire detector;

  iv   any storage encroaches within 300 mm of ceilings;

  v   a clear space of 500 mm is maintained below each automatic fire detector, and that the ability of the detector to receive the stimulus for which it has been designed to detect has not been impeded by other means;

  vi   any changes to the use or occupancy of an area make the existing types of automatic fire detector unsuitable for detection of fire or prone to unwanted alarms;

  vii   any building alterations or extensions require additional fire detection and alarm equipment to be installed.

c   The records of false alarms is checked. The rate of false alarms during the previous 12 months should be recorded and action recommended.

d   The standby battery is disconnected and full load alarm simulated.

e   Batteries and their connections are examined and momentarily load tested with the mains disconnected (other than those within devices such as manual call points, detectors and fire alarm sounders of a radio-linked system), to ensure that they are in good serviceable condition and not likely to fail before the next service visit. Vented batteries are examined to check that the specific gravity of each cell is correct.

f   The fire alarm functions of the control and indicating equipment are checked by the operation of at least one detector or manual call point on each circuit. An entry is made in the log book indicating which initiating devices have been used for these tests.

g   All controls and visual indicators at control and indicating equipment are checked for correct operation.

h   The operation of any facility for automatic transmission of alarm signals to an alarm receiving centre is checked. Where more than one form of alarm signal can be transmitted (e.g. fire and fault signals), the correct transmission of each signal should be confirmed.

i   All ancillary functions of the control and indicating equipment are tested.

j   All fault indicators and their circuits are checked, where practicable, by simulation of fault conditions.

k   All printers are tested to check if they operate correctly and that characters are legible. It should be ensured that all printer consumables are sufficient in quantity or condition to ensure that the printer can be expected to operate until the time of the next service visit.

7

l Radio systems of all types are to be serviced in accordance with the recommendations of the manufacturer.

m All further checks and tests recommended by the manufacturer of the control and indicating equipment and other components of the system are to be carried out.

n On completion of the work, any outstanding defects are reported to the responsible person, the system log book should be completed and a servicing certificate should be issued.

7.14.6 Twelve-monthly professional inspections and tests <small>(BS 5839-1 Clause 45.4)</small>

In addition to the six-monthly work recommended in Section 7.14.5 above, the following work should be carried out every year.

Note 1: The work described may be carried out over the course of two six-monthly service visits.

a The switch mechanism of every manual call point is tested, either by removal of a frangible element, insertion of a test key or operation of the device as it would be operated in the event of fire.

b All automatic fire detectors are examined to confirm that they have not been damaged, painted, or otherwise adversely affected. Then every detector is functionally tested to prove only that the detectors are connected to the system, are operational and are capable of responding to the phenomena they are designed to detect.

c Every heat detector is functionally tested by means of a suitable heat source, unless operation of the detector in this manner would then necessitate replacement of part or all of the sensing element (e.g. as in fusible link point detectors or non-integrating line detectors). Special test arrangements will be required for fusible link heat detectors. The heat source should not have the potential to ignite a fire; live flame should not be used, and special equipment might be necessary in explosive atmospheres.

d Point smoke detectors are to be functionally tested by a method that confirms that smoke can enter the detector chamber and produce a fire alarm signal (e.g. by use of apparatus that generates simulated smoke or suitable aerosols around the detector). It should be ensured that the material used does not cause damage to, or affect the subsequent performance of, the detector; the manufacturer's guidance on suitable materials should be followed.

e Optical beam smoke detectors are to be functionally tested by introducing signal attenuation between the transmitter and receiver, either by use of an optical filter, smoke or simulated smoke.

f Aspirating fire detection systems are to be functionally tested with each sampling point, or group of sampling points, in the pipework of the system treated as a point detector. Note that not all test products may be appropriate for the purpose.

g Carbon monoxide fire detectors are functionally tested by a method that confirms that carbon monoxide can enter the detector chamber and produce a fire alarm signal (e.g. by use of apparatus that generates carbon monoxide or a gas that has a similar effect as carbon monoxide on the electro-chemical cell).

WARNING: Carbon monoxide is a highly toxic gas, and suitable precautions should be taken in its use.

Note 2: It should be ensured that any test gas used does not cause damage to, or affect the subsequent performance of, the detector; the manufacturer's guidance on suitable test gases should be followed.

h Flame detectors are to be functionally tested by a method that confirms that the detector will respond to a suitable frequency of radiation and produce a fire alarm signal. The guidance of the manufacturer on testing of detectors should be followed.

i In fire detection systems that enable analogue values to be determined at the control and indicating equipment, it is confirmed that each analogue value is within the range specified by the manufacturer.

j Multi-sensor detectors should be operated by a method that confirms that products of combustion in the vicinity of the detector can reach the sensors and that a fire signal can be produced as appropriate. The guidance of the manufacturer on the manner in which the detector can be functionally tested effectively should be followed.

k All fire alarm devices are checked for correct operation. It should be confirmed that visual fire alarm devices are not obstructed from view and that their lenses are clean.

l All unmonitored, permanently illuminated filament lamp indicators at control and indicating equipment should be replaced.

m Radio signal strengths in radio-linked systems should be checked for adequacy.

n A visual inspection should be undertaken to confirm that all readily accessible cable fixings are secure and undamaged.

o The cause-and-effect programme should be confirmed as being correct.

p The standby power supply capacity is to be checked to establish it remains suitable for continued service.

q All further annual checks and tests recommended by the manufacturer of the control and indicating equipment and other components of the system should be carried out.

On completion of the work, any outstanding defects are to be reported to the responsible person and a record of the inspection and test should be made on the servicing certificate.

Note 3: Since stimulus of the sensing element through introduction of the phenomena or surrogate phenomena which the above detectors are designed to detect forms part of the test, use of a test button or a test magnet (for example) or compliance with item **i** above does not satisfy the recommendations given.

BS 7671:2008 Requirements for electrical installations

8

8.1 Chapter 56 Safety services

BS 5839-1 requires that BS 7671 be complied with (Clauses 29.1, 37.1, 37.2 and for testing 39.2). Chapter 56 'Safety services' of BS 7671 applies particularly to:

a emergency lighting
b fire detection and alarm systems
c fire evacuation systems
d fire pumps, fire rescue service lifts, CO_2 detection and alarm systems, fire ventilation systems, fire service communication systems.

Most of the requirements are common to both standards including:

a safety circuits to be independent of and unaffected by other circuits (regulation 560.7)
b switchgear to be identified and in locations accessible only to skilled or instructed persons
c cables required to operate in fire conditions to be:
  i Fire-resistant, complying with BS EN 50362 or BS EN 50200, appropriate for the cable size, and with BS EN 60332-1-2.
  Note: BS EN 50200 is called up in BS 5839-1 with the necessary specific performance requirement (PH 60 or PH 120) missing from BS 7671.
  Note: BS EN 60332-1-2:2004: *Tests on electric and optical fibre cables under fire conditions. Test for vertical flame propagation for a single insulated wire or cable. Procedure for 1 kW pre-mixed flame* is complied with by most cables other than polyethylene cables and some rubber.
  ii Able to maintain the necessary fire and mechanical protection.
  Note: BS 5839-1 Clause 26.2 specifies cables to BS EN 60702-1, BS 7629 and BS 7846 as suitable.

There are in practice no additional cable performance requirements in Chapter 56 of BS 7671 to those in BS 5839-1.

8.2 Chapter 52 Selection and erection of wiring systems

Regulations 522.6.6 and 522.6.7 of BS 7671 are copied below as they are particularly relevant to fire systems and much debated, in particular regulation 522.6.7. They are presumed to apply to all low-voltage cables (not exceeding 1000 V a.c. or 1500 V d.c.), other than ELV (extra low voltage) cables (not exceeding 50 V a.c. or 120 V ripple-free d.c.).

522.6.6 A cable concealed in a wall or partition at a depth of less than 50 mm from a surface of the wall or partition shall:

(i) incorporate an earthed metallic covering which complies with the requirements of these Regulations for a protective conductor of the circuit concerned, the cable complying with BS 5467, BS 6346, BS 6724, BS 7846, BS EN 60702-1 or BS 8436, or

(ii) be enclosed in earthed conduit complying with BS EN 61386 and satisfying the requirements of these Regulations for a protective conductor, or

(iii) be enclosed in earthed trunking or ducting complying with BS EN 50085 and satisfying the requirements of these Regulations for a protective conductor, or

(iv) be mechanically protected against damage sufficient to prevent penetration of the cable by nails, screws and the like, or

(v) be installed in a zone within 150 mm from the top of the wall or partition or within 150 mm of an angle formed by two adjoining walls or partitions. Where the cable is connected to a point, accessory or switchgear on any surface of the wall or partition, the cable may be installed in a zone either horizontally or vertically, to the point, accessory or switchgear. Where the location of the accessory, point or switchgear can be determined from the reverse side, a zone formed on one side of a wall of 100 mm thickness or less or partition of 100 mm thickness or less extends to the reverse side.

522.6.7 Where Regulation 522.6.6 applies and the installation is not intended to be under the supervision of a skilled or instructed person, a cable installed in accordance with Regulation 522.6.6 (v), and not complying with Regulation 522.6.6 (i), (ii), (iii) or (iv), shall be provided with additional protection by means of an RCD having the characteristics specified in Regulation 415.1.1.

Regulations 522.6.5 and 522.6.8 have similar requirements for cables in floors and ceilings at a depth of less than 50 mm and in walls and partitions with metal parts and frames.

Clause 25.2h) of BS 5839-1 requires that 'The circuit supplying the fire alarm system should not be protected by a residual current device unless this is necessary to comply with the requirements of BS 7671', this is understood to mean such circumstances as TT supplies where there is practically no alternative to the use of an RCD to provide for safety, see also Clause 26.1.

Most commercial and industrial premises will be 'under the supervision of a skilled or instructed person' as the Electricity at Work Regulations will require the employer to take precautions to provide for the safety of people using the building. This includes enforcing rules to ensure the drilling etc. of walls or partitions (that may lead to cable damage) will only be carried out by competent persons. In these circumstances BS 7671 does not require RCD protection.

If this cannot be ensured cables will have to be either:

▶ installed on the surface
▶ incorporate an earthed metallic covering such as cables to
    BS 6724 (armoured low smoke),
    BS 7846 (armoured fire resistant),
    BS EN 60702-1 (mineral insulated) or
    BS 8436 (screened low smoke and gases).

8.3   Section 422.2 Conditions for evacuation in an emergency

To provide for safety of persons evacuating a building in an emergency, wiring systems installed in escape routes with conditions BD2, BD3 or BD4 (see Table 8.1) must be non-flame propagating.

▼ **Table 8.1**   Evacuation condition categories

Code	Occupation density	Evacuation conditions	Example
BD2	Low	Difficult	High-rise buildings not open to the public
BD3	High	Easy	Locations open to the public (theatres, cinemas, department stores, shopping centres)
BD4	High	Difficult	High-rise buildings open to the public

Cables are required to comply with BS EN 50266 and BS EN 61034-2 – see Table 8.2.

Where conduit and trunking systems are used they must be to the British Standard and be classified as fire resistant by these Standards.

▶ For conduits the Standard referred to is BS EN 61386-1.
▶ For trunking the Standard referred to is BS EN 50085.

▼ **Table 8.2** Cable fire performance requirements from Section 422 of BS 7671

Performance requirement	Regulation	Location	Suitable cables*
BS EN 50266[a]	422.2.1	Escape routes	BS 6724[d]
	422.3.4	Flammable stored/processed materials and where cables bunched, or long vertical runs	BS 7846[e] BS 7835[f]
	527	Particular risks	
BS EN 61034-2[b] (was BS EN 50268-2, was BS 7622-2)	422.2.1	Escape routes	BS 6724[d]
	711.521	Exhibitions and shows where no fire alarm system	BS 7846[e] BS 7835[f]
BS EN 60332-1-2[c]	422.3.4	Flammable stored/processed materials	Most cables meet this requirement other than polyethylene sheath cables and some rubber cables
	422.4.5	Combustible construction materials	

Notes:

* Only British Standards that require the cables to meet the stated performance have been listed. Manufactures may have tested cables to meet a higher performance than required by a BS and therefore may be suitable for use.

a BS EN 50266-1:2001 *Common test methods for cables under fire conditions. Test for vertical flame spread of vertically-mounted bunched wires or cables. Apparatus.*

b BS EN 61034-2:2005 *Measurement of smoke density of cables burning under defined conditions. Test procedure and requirements* (replaces BS EN 50268-2).

c BS EN 60332-1-2:2004 *Tests on electric and optical fibre cables under fire conditions. Test for vertical flame propagation for a single insulated wire or cable. Procedure for 1 kW pre-mixed flame.*

d BS 6724 *Armoured electric cables having thermosetting insulation and low emission of smoke and gases.*

e BS 7846 *Armoured fire resistant electric cables having thermosetting insulation and low emission of smoke and gases.*

f BS 7835 *Armoured cables with thermosetting insulation for rated voltages from 3.8/6.6 kV to 19/33 kV having low emission of smoke and corrosive gases when affected by fire.*

8.4 Protection against overload

BS 7671 allows protection against overload to be omitted (regulations 433.3.3 and 560.7.3) to provide for a more reliable supply. This allows overcurent devices (fuses and circuit-breakers to be 'oversized' so that unwanted operation is avoided. As the electrical load of the fire system will not change (without modification), this is within the general rules. Fault protection and shock protection must be provided.

8.5　Section 527.2 Sealing of wiring system penetrations

8.5.1　External sealing

BS 7671 requires that where a wiring system (cables, conduit, trunking etc.) passes through elements of building construction such as floors, walls, roofs, ceilings, partitions or cavity barriers, the openings remaining after passage of the wiring system shall be sealed according to the degree of fire-resistance (if any) prescribed for the respective element of building construction before penetration.

During the erection of a wiring system temporary sealing arrangements shall be provided as appropriate and during alteration work, sealing which has been disturbed shall be reinstated as soon as practicable.

8.5.2　Internal sealing

A wiring system (cables, conduit, trunking etc.) which penetrates elements of building construction having specified fire-resistance shall be internally sealed to the degree of fire-resistance of the respective element before penetration as well as being externally sealed.

However a wiring system classified as non-flame propagating according to the relevant product standard and having a maximum internal cross-sectional area of 710 mm^2 need not be internally sealed provided that:

i　the system satisfies the test of BS EN 60529 for IP33, and
ii　any termination of the system in one of the compartments, separated by the building construction being penetrated, satisfies the test of BS EN 60529 for IP33.

Sealing systems are required to be fire resistant, structurally adequate and prevent ingress of water where appropriate and generally fit for purpose.

British Standards

9

9.1 Introduction

Types of British Standard are defined below.

BS (British Standard)

There are five types of British Standard: Specifications, Methods, Guides, Vocabularies and Codes of Practice. These all carry the prefix BS and all have the same status and authority.

PD (Published document)

This is a catch-all category for standards-type documents that do not have the same status as a BS. Some PDs are adoptions of CEN, CENELEC, ISO or IEC publications that are themselves not standards (e.g. Technical Reports). Others are derived from British Standards that conflict with ENs but are still needed by industry (e.g. PD 5500: *Pressure vessels*). Others are developed by national committees but for one reason or another do not go through the whole development process in strict conformity with BS.

BS EN

This is a European Standard. As a member of CEN and CENELEC, the British Standards Institution (BSI) is obliged to adopt all European Standards and to withdraw any national standards that might conflict with them (this is why we sometimes change the status of a standard from BS to PD if we want to keep it on the market). They are published in the UK as BS EN.

PAS (Publicly available specification)

A PAS can be seen as a step in the process of standardisation. It includes useful and practical information that can be made available quickly to suit the market needs of the developers and users of a product, process or service.

A full standard requires several more stages of development before full consensus is achieved. The rationale for publishing a PAS is that while it may not have the full breadth of agreement of a standard, the speed of delivery and the high-calibre quality of the content enables users of the PAS to reap significant benefit.

© The Institution of Engineering and Technology

9.1.1 Standards referenced in the Building Regulations

There are three main groups of fire system standards referenced in the Building Regulations:

a BS 5839 *Fire detection and alarm systems for buildings*
b BS 5588 *Fire precautions in the design, construction and use of buildings*
c BS 7974 *Application of fire safety engineering principles to the design of buildings*

There are many other British Standards referred to in the above British Standards, in particular BS 5446 *Fire detection and fire alarm devices for dwellings.*

9.2 BS 5839 Fire detection and alarm systems for buildings

This standard is concerned with the design and installation of equipment for the fire detection and fire alarm system – that is, the electrical fire detection and alarm system. It is not concerned with fire provisions taken in the design and construction of the building itself.

BS 5839-1 *Fire detection and fire alarm systems for buildings. Code of practice for system design, installation, commissioning and maintenance* is the subject of Chapter 7. This is an important standard, as it leads to compliance of the fire detection and alarm system requirements of the Building Regulations. Its scope is all buildings including dwellings. It is referenced in both Volume 1 and Volume 2 of Approved Document B (see Table 7.1).

BS 5839-6 *Fire detection and fire alarm systems for buildings. Code of practice for the design, installation and maintenance of fire detection and fire alarm systems **in dwellings*** is the subject of Chapter 5 and is concerned only with dwellings.

This standard is referenced in Volume 1 (dwellinghouses) of Approved Document B: Fire safety. The Approved Document specifies the particular category that will lead to compliance, but Approved Document B requires a further additional smoke alarm in the principal/largest bedroom, see paragraph 4.5.1. A full list of the current parts of BS 5839 is listed in Table 9.1. The series includes design codes of practice and equipment standards.

▼ **Table 9.1**   BS 5839 Fire detection and fire alarm systems for buildings

Type of standard	Standard number	Title
Code of practice	BS 5839-1:2002+A2 2008	Fire detection and fire alarm systems for buildings. Code of practice for system design, installation, commissioning and maintenance
Specification	BS 5839-3:1988	Fire detection and fire alarm systems for buildings. Specification for automatic release mechanisms for certain fire protection equipment
Code of practice	BS 5839-6:2004	Fire detection and fire alarm systems for buildings. Code of practice for the design, installation and maintenance of fire detection and fire alarm systems in dwellings
Code of practice	BS 5839-8:2008	Fire detection and fire alarm systems for buildings. Code of practice for the design, installation, commissioning and maintenance of voice alarm systems
Code of practice	BS 5839-9:2003	Fire detection and fire alarm systems for buildings. Code of practice for the design, installation, commissioning and maintenance of emergency voice communication systems

9.3   BS 9999:2008 Code of Practice for fire safety in the design, management and use of buildings

This standard provides an advanced structured approach to fire safety using risk-based design. It provides a link between Approved Document B and BS 7974.

The standard is applicable to the design of new buildings, and to alterations, extensions and changes of use of an existing building, with the exception of individual homes and with limited applicability in the case of certain specialist buildings. It also provides guidance on the ongoing management of fire safety in a building throughout the entire life cycle of the building, including guidance for designers to ensure that the overall design of a building assists and enhances the management of fire safety. It can be used as a tool for assessing existing buildings, although fundamental change in line with the guidelines might well be limited or not practicable.

9.4   BS 5588 Fire precautions in the design, construction and use of buildings

Note: All parts except BS 5588-1 (see Table 9.2) have been superseded by BS 9999.

▼ **Table 9.2**   BS 5588 Fire precautions in the design, construction and use of buildings

Type of standard	Standard number	Title
Code of practice	BS 5588-1:1990	Fire precautions in the design, construction and use of buildings. Code of practice for residential buildings

9.5   BS 7974 Application of fire safety engineering principles to the design of buildings

Reference is made to this standard in the general introduction to both Volumes 1 and 2 of Approved Document B as follows:

Fire safety engineering

Fire safety engineering can provide an alternative approach to fire safety. It may be the only practical way to achieve a satisfactory standard of fire safety in large and complex buildings. Fire safety engineering may also be suitable for solving a problem with an aspect of the building design which otherwise follows the provisions in this document.

British Standard BS 7974:2001 *Application of fire safety engineering principles to the design of buildings* and supporting published documents provide a framework and guidance on the design and assessment of fire safety measures in buildings. Following the discipline of BS 7974 should enable designers and building control bodies to be aware of the relevant issues, the need to consider the complete fire safety system, and to follow a disciplined analytical framework.

This standard is not discussed in this Guide, as electrical designers are generally not involved in the early stages of building design.

A full list of the parts of BS 7974 is in Table 9.3.

▼ **Table 9.3**   BS 7974 Application of fire safety engineering principles to the design of buildings

Type of standard	Standard number	Title
Code of practice	BS 7974:2001	Application of fire safety engineering principles to the design of buildings. Code of practice
Published document	PD 7974-0:2002	Application of fire safety engineering principles to the design of buildings. Guide to design framework and fire safety engineering procedures
Published document	PD 7974-1:2003	Application of fire safety engineering principles to the design of buildings. Initiation and development of fire within the enclosure of origin (Sub-system 1)
Published document	PD 7974-2:2002	Application of fire safety engineering principles to the design of buildings. Spread of smoke and toxic gases within and beyond the enclosure of origin (Sub-system 2)
Published document	PD 7974-3:2003	Application of fire safety engineering principles to the design of buildings. Structural response and fire spread beyond the enclosure of origin (Sub-system 3)
Published document	PD 7974-4:2003	Application of fire safety engineering principles to the design of buildings. Detection of fire and activation of fire protection systems (Sub-system 4)
Published document	PD 7974-5:2002	Application of fire safety engineering principles to the design of buildings. Fire service intervention (Sub-system 5)
Published document	PD 7974-6:2004	The application of fire safety engineering principles to fire safety design of buildings. Human factors. Life safety strategies. Occupant evacuation, behaviour and condition (Sub-system 6)
Published document	PD 7974-7:2003	Application of fire safety engineering principles to the design of buildings. Probabilistic risk assessment

9

9.6   Equipment standards

9.6.1   BS 5446 Fire detection and fire alarm devices for dwellings

This standard is concerned with equipment specifications: smoke alarms, heat alarms and alarm kits for the deaf and hard of hearing. The various parts are listed in Table 9.4.

▼ **Table 9.4**   BS 5446 Fire detection and fire alarm devices for dwellings

Type of standard	Standard number	Title
Standard	BS 5446-1:2000	Fire detection and fire alarm devices for dwellings. Specification for smoke alarms
Standard	BS 5446-2:2003	Fire detection and fire alarm devices for dwellings. Specification for heat alarms
Standard	BS 5446-3:2005	Fire detection and fire alarm devices for dwellings. Specification for smoke alarm kits for deaf and hard of hearing people

9.6.2   BS EN 54 Fire detection and fire alarm systems

The various parts of BS EN 54 are listed in Table 9.5.

▼ **Table 9.5**   BS EN 54 Fire detection and fire alarm systems

Type of standard	Standard number	Title
Standard	BS EN 54-1:1996	Fire detection and fire alarm systems. Introduction
Standard	BS EN 54-2:1998	Fire detection and fire alarm systems. Control and indicating equipment
Standard	BS EN 54-3:2001	Fire detection and fire alarm systems. Fire alarm devices. Sounders
Standard	BS EN 54-4:1998	Fire detection and fire alarm systems. Power supply equipment
Standard	BS EN 54-5:2001	Fire detection and fire alarm systems. Heat detectors. Point detectors
Standard	BS EN 54-7:2001	Fire detection and fire alarm systems. Smoke detectors. Point detectors using scattered light, transmitted light or ionization
Standard	BS EN 54-10:2002	Fire detection and fire alarm systems. Flame detectors. Point detectors
Standard	BS EN 54-11:2001	Fire detection and fire alarm systems. Manual call points
Standard	BS EN 54-12:2002	Fire detection and fire alarm systems. Smoke detectors. Line detectors using an optical light beam
Standard	BS EN 54-13:2005	Fire detection and fire alarm systems. Compatibility assessment of system components

▼ **Table 9.5**   *continued*

Type of standard	Standard number	Title
Standard	BS EN 54-17:2005	Fire detection and fire alarm systems. Short-circuit isolators
Standard	BS EN 54-18:2005	Fire detection and fire alarm systems. Input/output devices
Standard	BS EN 54-20:2006	Fire detection and fire alarm systems. Aspirating smoke detectors
Standard	BS EN 54-25:2008	Fire detection and fire alarm systems. Components using radio links

9.6.3   BS 7346 Components for smoke and heat control

The various parts of BS 7346 are listed in Table 9.6.

▼ **Table 9.6**   BS 7346 Components for smoke and heat control

Type of standard	Standard number	Title
Code of practice	BS 7346-5	Functional recommendations and calculation methods for smoke and heat exhaust ventilation systems, employing time-dependent design fires
Standard	BS 7346-6	Specifications for cable systems

9.7   BS 7273 Code of practice for the operation of fire protection measures

The various parts of BS 7273 are listed in Table 9.7.

▼ **Table 9.7**   BS 7273 Code of practice for the operation of fire protection measures

Type of standard	Standard number	Title
Code of practice	BS 7273-1:2000	Code of practice for the operation of fire protection measures. Electrical actuation of gaseous total flooding extinguishing systems
Code of practice	BS 7273-2:1992	Code of practice for the operation of fire protection measures. Mechanical actuation of gaseous total flooding and local application extinguishing systems
Code of practice	BS 7273-3:2000	Code of practice for the operation of fire protection measures. Electrical actuation of pre-action sprinkler systems
Code of practice	BS 7273-4:2007	Code of practice for the operation of fire protection measures. Actuation of release mechanisms for doors

9.8   Test standards

The two relevant test standards are listed in Table 9.8.

▼ **Table 9.8**   Test standards

Type of standard	Standard number	Title
Standard	BS EN 50200:2000	Method of test for resistance to fire of unprotected small cables for use in emergency circuits
Standard	BS 8434-1:2003	Methods of test for assessment of the fire integrity of electric cables – Part 1: Test for unprotected small cables for use in emergency circuits – BS EN 50200 with addition of water

Scottish building standards 10

10.1 Introduction

This chapter explains the requirements of the Building (Scotland) Act 2003 and associated legislation. Detailed information on the Scottish system, including building regulations, can be found at the Building Standards Division (BSD) of the Directorate for the Built Environment website: www.sbsa.gov.uk.

10.2 Legislation

10.2.1 The Building (Scotland) Act 2003 (2003 asp 8)

This Act gives Scottish Ministers the power to make building regulations to secure the health, safety, welfare and convenience of persons in or about buildings and of others who may be affected by buildings or matters connected with buildings; to further the conservation of fuel and power; and to further the achievement of sustainable development.

10.2.2 The Building (Scotland) Regulations 2004

The Building (Scotland) Regulations 2004 are made under the powers of the Building (Scotland) Act 2003 and apply solely in Scotland.

The regulations apply to construction, alteration, conversion and demolition of buildings and also to the provision of services, fittings and equipment in or in connection with buildings except where they are specifically exempted from regulations 8 to 12 (Schedule 1 to regulation 3). Schedule 3 to regulation 5 lists work that must comply but with certain exceptions does not require a building warrant.

The regulations prescribe functional standards for buildings, which can be found in Schedule 5 to regulation 9. The regulations are amended periodically. However, it is the regulations in force at the time of the application that must be complied with.

10.2.3 Responsibility for compliance with building regulations

In Scotland, final responsibility for compliance with building regulations rests with the 'relevant person', (defined in section 17 of the Act) who will normally be the owner or developer of a building. However, any person carrying out work including electrical work has a duty to ensure their work complies with building regulations.

10.3 Scottish building standards guidance documents

10.3.1 Technical guidance

The Scottish Building Standards Technical Handbooks are published in two volumes: domestic buildings and non-domestic buildings. Each Handbook is split into seven sections. Section 0 covers general issues and offers an introduction and guidance to the Building Regulations. Sections 1 to 6 include the functional standards together with guidance on how to comply with them; they are summarised in the table below.

section 1	structure	section 4	safety
section 2	fire	section 5	noise
section 3	environment	section 6	energy

Section 2 of both the domestic and the non-domestic volume is of particular interest to persons installing fire detection and alarm systems. Section 2.11 of both volumes is reproduced below.

standard **2.11** mandatory	Every *building* must be designed and *constructed* in such a way that in the event of an outbreak of fire within the *building*, the occupants are alerted to the outbreak of fire. **Limitation:** This standard applies only to a *building* which: (a) is a *dwelling*; (b) is a *residential building*; or (c) is an enclosed shopping centre.

For example the guidance in 2.11.1 for dwellings with no storey more than 200 m^2 is as follows:

A *dwelling* where no *storey* is more than 200 m^2 should be provided with 1 or more *smoke alarms* located on each *storey* with a standby supply to BS 5446: Part 1: 2000 and installed in accordance with the guidance in clause 2.11.2.

Dwellings with a storey greater than 200 m^2 are required to be provided with a fire detection and fire alarm system in accordance with BS 5839-6:2004 for a Grade C Category LD2 system.

10.3.2 Procedural guidance

Scottish Ministers have published a Procedural Handbook which explains in more detail the Scottish system and is accessible on the BSD website. Unlike the Technical Handbooks, this Handbook has no specific legal status, but is designed to clarify the procedures.

See http://www.sbsa.gov.uk/proced_legislation/proc_handbook.htm

10.3.3 Certification guidance

Scottish Ministers have published a Certification Handbook for schemes under section 7(2) of the Building Scotland Act 2003 and is accessible on the BSD website. The Handbook provides guidance on the optional procedure using Approved Certifiers of Construction (see section 10.4.3).

See http://www.sbsa.gov.uk/tech_handbooks

10.3.4 Practical guidance

Practical fire safety guidance for compliance with the Fire (Scotland) Act is found on the Scottish Government Firelaw website: www.infoscotland.com/firelaw.

A series of sector-specific guides can be downloaded providing practical fire safety guidance for those with responsibilities under Part 3 of the Fire (Scotland) Act 2005, as amended, and the Fire Safety (Scotland) Regulations 2006. Part of the explanatory text from the website is reproduced below.

> Where possible, the guides do not set down prescriptive standards, but provide recommendations regarding the fire safety risk assessment process, the reduction of risk and guidance on fire safety measures that can be implemented to mitigate risk. However, there is no obligation to adopt any particular solution in the guides if the outcomes of a fire safety risk assessment can be met in some other way.
>
> Enforcing authorities are required to take into account the content of the guides to assist in determining whether enforcement action may be necessary but in doing so they should have a flexible approach to enforcement.

The guides listed below are for duty holders, e.g. employers:

a   Care Homes Guide
b   Offices, Shops & Similar Premises Guide
c   Factories & Storage Premises Guide
d   Educational & Day Care for Children Premises Guide
e   Small Premises Providing Sleeping Accommodation Guide
f   Medium & Large Premises Providing Sleeping Accommodation
g   Transport Premises Guides
h   Places of Entertainment and Assembly Guide
i   Healthcare Premises Guide

10.4 Procedures

In Scotland, a building warrant is required for building work unless:

a it is not a 'building' as defined in the Building (Scotland) Act 2003
b it is exempt from building regulations (Schedule 1 to regulation 3)
c it does not require a warrant but still must comply with building regulations (Schedule 3 to regulation 5).

Anyone intending to carry out electrical work that requires a warrant should note that:

a a building warrant must be obtained before work starts
b a completion certificate must be submitted when work is complete
c a completion certificate must be accepted by the Verifier before occupation or use of a new building (or extension) is permitted.

10.4.1 Work requiring a building warrant

Except where described above, a building warrant must be applied for and be granted by the Verifier before work can commence. It is an offence to commence work without a warrant. The application is made to the Verifier, i.e. the local authority building standards department (currently only the 32 local authorities have been appointed as Verifiers for their geographical area) who will assess the application and issue the building warrant if proposals are considered to comply with building regulations.

The use of an Approved Certifier of Design or Construction (see section 10.4.3 below) does not obviate the need to obtain a building warrant but the certificate they issue must be accepted by the Verifier. If an Approved Certifier of Design is used, the warrant fee can be discounted and if an Approved Certifier of Construction is used there may be a partial refund.

The Verifier must be notified when work commences. Notification is usually made by the relevant person (normally the owner or developer of the building) or his agent. Where it is intended to use an Approved Certifier of Construction, such as for electrical installations, this should be notified to the Verifier prior to the electrical work commencing.

Designers and installers are directed to section 2 'Fire' of the Scottish Building Standards Technical Handbooks for detailed guidance on issues including openings and fire stopping (standards 2.2 and 2.4), escape route lighting (standard 2.10), fire alarm and detection systems (standard 2.11) and automatic life safety fire suppression systems (standard 2.15). A comparison between Scottish and English guidance documents is included in Section 10.7.

10.4.2 Completion Certificate

If a building warrant has been granted a Completion Certificate, it must be submitted to the Verifier once works are complete. This states that works have been completed in accordance with building regulations and the building warrant. If the Verifier, after making their reasonable enquiries, is satisfied, they must accept the Completion Certificate. If it is rejected, the Verifier will identify the reason for rejection.

Where an Approved Certifier of Construction for electrical installations is used, they will issue a Certificate of Construction (Electrical Installations to BS 7671) which the Verifier must accept as compliance for the specific work described. It should be submitted with the Completion Certificate.

10.4.3 Approved Certifiers of Design/Approved Certifiers of Construction

The Building (Scotland) Act 2003 establishes a role for suitably qualified people, businesses or other bodies, when appointed by the Scottish Ministers, to certify that certain design or construction work complies with the Building Regulations. Two roles are designated – Approved Certifiers of Design and Approved Certifiers of Construction – both of which certify compliance with the Building Regulations, as laid down in the scope of the certification scheme run by the scheme provider. Further information on certification schemes may be found on the BSD website.

Approved Certifiers of Construction are responsible for the construction or installation of specified parts of a building, such as the electrical installation, and must have due regard for compliance with the full range of building standards, not just those applicable to their particular work.

The approved schemes for the Certification of Design or Construction are found on the BSD website: www.sbsa.gov.uk/current_standards/CurrentSchemes.

10.4.4 Buildings and services exempt from building regulations

Certain buildings and services are exempt from regulations 8 to 12 and a building warrant is not required. They are set out in Schedule 1 to regulation 3 of the Building (Scotland) Regulations and include exceptions where the Building Regulations do apply. The exempt types include:

a buildings or work covered by other legislation
b buildings or work not frequented by people
c specialised buildings or work where application of the regulations is largely inappropriate
d buildings or works that are minor where enforcement is not in the public interest
e temporary buildings or works.

10.4.5 Works not requiring building warrant

Certain works subject to building regulations do not require a building warrant but works must still meet the Building Regulations. They are set out in Schedule 3 to regulation 5 of the Building (Scotland) Regulations and include exceptions where a warrant is required. For example:

a Any work to or in a house with a storey height not exceeding 4.5 m but only if the exceptions do not apply (type 1).
b Any work to or in a non-residential building to which members of the public do not have access, with a storey height not exceeding 7.5 m but only if the exceptions do not apply (type 2).

In both cases the exceptions include, but are not limited to, structural work, work to a separating wall and alterations to roofs or external walls.

There are other specific types of work that can be done to any building that also do not require a warrant (types 3 to 26).

Two matrices identifying electrical work not requiring a warrant have been prepared jointly by the Scottish Association of Building Standards Managers and the BSD. They cover domestic and non-domestic work and are accessible on the BSD website.

10.5   Electrical installations

10.5.1   Standard 4.5 Electrical safety

Electrical installations must comply with standard 4.5 (electrical safety) which states that every building must be designed and constructed in such a way that the electrical installation does not threaten the health and safety of the people in and around the building, and does not become a source of fire.

Guidance to standard 4.5 cites BS 7671 as a means of complying with the functional standard and addresses, in brief, general low-voltage installations, extra low-voltage installations, installations operating above low voltage and socket outlets in bathrooms and rooms containing a shower.

10.5.2   Standard 4.6 Electrical fixtures

Electrical installations must comply with standard 4.6 (electrical fixtures) which states that every domestic building must be designed and constructed in such a way that electric lighting points and socket outlets are provided. Minimum numbers of lighting points and socket outlets are given.

Guidance to standard 4.6 includes for lighting, light switches in common areas, entry phone systems and socket outlets. The recommended positioning of electrical fixtures, including heights of switches and socket outlets, is given in standard 4.8.

10.5.3   Fitness and durability of materials, workmanship and access for maintenance

Regulation 8 requires that materials, fittings and components used should be suitable for their purpose, correctly used or applied, and sufficiently durable, taking account of normal maintenance practices, to meet the requirements of these regulations. It also implements the intention of the Construction Products Directive, that specification of construction products should not be used to effectively bar the use of construction products or processes from other European countries. The relevant countries are those in the European Union, and those who in the European Economic Area Act of 1993 agreed to adopt the same standards.

10.6 Guidance on electrical work not requiring a warrant

10.6.1 Domestic buildings

Building (Scotland) Regulations 2004
Regulation 5, Schedule 3

Guidance on electrical work
not requiring a warrant

SABSM

DOMESTIC BUILDINGS	WORK TO EXISTING BUILDINGS			
	Type[1]	Flat	House (up to 2 storeys)	House (3 storeys & above)
Repairs and replacement				
Re-wiring[2]	24	required	not required	required
Electrical fixtures, e.g. luminaries	24	not required	not required	not required
New work				
Electrical work affected by demolition or alteration of the roof, external walls or elements of structure	1	required	required	required
Electrical work adversely affecting a separating wall, e.g. recessed sockets	1	required	required	required
New power socket outlets	1	required	not required	required
Mains operated fire alarm system	1	required	not required	required
Electrical work to automatic opening ventilators (including auto-detection)	1	required	not required	required
Electrically operated locks	1	required	not required	required
Wiring to artificial lighting	1	required	not required	required
Wiring to emergency lighting	1	required	not required	required
Electrical work associated with sprinkler system	1	required	not required	required
Electrical work associated with new boiler (large)	1	required	not required	required
Electrical work associated with new boiler (small)	6	not required	not required	not required
Electrical work associated with new shower	11, 12	not required	not required	not required
Electrical work associated with new extractor fan	13	not required	not required	not required
Extra low-voltage installations	22	not required	not required	not required

Notes:
1 Building work type as referenced in schedule 3.
2 A building warrant is not required for rewiring where it is a repair or replacement works to a level equal to the installation (or part thereof) being repaired or replaced.

10.6.2 Non-domestic buildings

**Building (Scotland) Regulations 2004
Regulation 5, Schedule 3**

Guidance on electrical work
not requiring a warrant

SABSM

NON-DOMESTIC BUILDINGS	WORK TO EXISTING BUILDINGS			
	Non-residential buildings with a storey, or creating a storey, not more than 7.5 m			Other non-domestic buildings
	Type[1]	No public access[2]	Public access	
Repairs and replacement				
Re-wiring[3]	24	not required	required	required
New work				
Electrical work affected by demolition or alteration of the roof, external walls or elements of structure	2	required	required	required
Electrical work adversely affecting a separating wall, e.g. recessed sockets	2	required	required	required
Electrical work adversely affecting a loadbearing wall	2	required	required	required
New power socket outlets	2	not required	required	required
Automatic fire detection system	2	not required	required	required
Electrical work to automatic opening ventilators	2	not required	required	required
Electrical work to automatic fire dampers	2	not required	required	required
Electrically operated locks	2	not required	required	required
Wiring to artificial lighting	2	not required	required	required
Wiring to emergency lighting	2	not required	required	required
Outdoor luminous tube signs[4]	2	not required	not required	not required
Electrical work associated with new boiler (large)	2	not required	required	required
Electrical work associated with new boiler (small)	6	not required	not required	not required
Electrical work associated with new shower	11,12	not required	not required	not required
Electrical work associated with new extract fan	13	not required	not required	not required
Extra low voltage installations	22	not required	not required	not required

Notes:
1   Building work type as referenced in schedule 3.
2   Non-residential buildings to which the public does not have access may include:
   •   Existing offices
   •   Existing storage buildings

- Existing industrial buildings, e.g. factories and workshops
- Existing assembly and entertainment buildings not open to the public, e.g. some educational buildings and private members clubs.

Non-residential buildings to which the public has access may include:

- Existing assembly and entertainment buildings open to the public, e.g. community schools, pubs and clubs.

3   A building warrant is not required for rewiring where it is a repair or replacement works to a level equal to the installation (or part thereof) being repaired or replaced.

4   Subject to the Town and Country Planning (Control of Advertisement) (Scotland) Regulations 1984.

10.7   Comparison between Scottish and English guidance documents

The following table lists the Approved Documents and their equivalents within the Scottish building standards system.

Scottish Technical Handbooks	England and Wales Approved Documents
Section 1 (structure)	A (Structure)
Section 2 (fire)	B (Fire safety)
Section 3 (environment) – standards 3.1 to 3.4 and 3.10 and 3.15	C (Site preparation and resistance to contaminants and moisture)
No equivalent	D (Toxic substances)
Section 5 (noise) – applies to dwellings only	E (Resistance to the passage of sound)
Section 3 (environment) – standards 3.14 (ventilation) and 3.10 (precipitation)	F (Ventilation)
Section 3 (environment) – standard 3.12 and section 4 (safety) – standard 4.9 (danger from heat)	G (Hygiene)
Section 3 (environment) – standards 3.5 to 3.9 (drainage) and 3.25 and 3.26 (waste storage)	H (Drainage and waste disposal)
Section 3 (environment) – standards 3.17 to 3.24 and section 4 (safety) – standard 4.11	J (Combustion appliances and fuel storage systems)
Section 4 (safety) - standards 4.3, 4.4, 4.8 and 4.12	K (Protection from falling, collision and impact)
Section 6 – (energy)	L (Conservation of fuel and power)
Section 3 (environment) – standards 3.11 and 3.12 and section 4 (safety) – standards 4.1, 4.2, 4.3, 4.6, 4.7 and 4.10	M (Access to and use of buildings)
Section 4 (Safety) – standard 4.8	N (Glazing – safety in relation to impact, opening and cleaning)
Section 4 (Safety) – standard 4.5	P (Electrical safety – dwellings)

In addition, the Technical Handbooks contain functional standards for which there are no direct equivalents in England and Wales. These are 'facilities in dwellings', 'heating' and 'natural lighting' (standards 3.11, 3.13 and 3.16).

Whilst the application of the principles in England and Wales are equally valid in Scotland, detailed recommendations within the BSD guidance may differ. Designers and installers are therefore advised to familiarise themselves with the standards and guidance within the Scottish Building Standards Technical Handbooks (domestic and non-domestic) prior to undertaking work in Scotland.

Index

Index

IEE Wiring Regulations and associated publications

The IEE prepares regulations for the safety of electrical installations for buildings, the *IEE Wiring Regulations* (BS 7671 *Requirements for Electrical Installations*), which have now become the standard for the UK and many other countries. It also recommends, internationally, the requirements for ships and offshore installations. The IEE provides guidance on the application of the installation regulations through publications focused on the various activities from design of the installation through to final test and then maintenance. This includes a series of eight Guidance Notes, two Codes of Practice and Model Forms for use in Wiring Installations.

Requirements for Electrical Installations BS 7671:2008 (IEE Wiring Regulations, 17th Edition)

Order book PWR1700B Paperback 2008
ISBN: 978-0-86341-844-0 **£75**

On-Site Guide (BS 7671:2008 17th Edition)

Order book PWGO170B 188pp Paperback 2008
ISBN: 978-0-86341-854-9 **£22**

Wiring Matters Magazine **FREE**

If you wish to receive a FREE copy or advertise in Wiring Matters please visit
www.theiet.org/wm

IEE Guidance Notes

A series of Guidance Notes has been issued, each of which enlarges upon and amplifies the particular requirements of a part of the IEE Wiring Regulations.

Guidance Note 1: Selection & Erection of Equipment, 5th Edition

Order book PWG1170B 216pp Paperback 2009
ISBN: 978-0-86341-855-6 **£30**

Guidance Note 2: Isolation & Switching, 5th Edition

Order book PWG2170B 88pp Paperback 2009
ISBN: 978-0-86341-856-3 **£25**

Guidance Note 3: Inspection & Testing, 5th Edition

Order book PWG3170B 128pp Paperback 2008
ISBN: 978-0-86341-857-0 **£25**

Guidance Note 4: Protection Against Fire, 5th Edition

Order book PWG4170B 104pp Paperback 2009
ISBN: 978-0-86341-858-7 **£25**

Guidance Note 5: Protection Against Electric Shock, 5th Edition

Order book PWG5170B 144pp Paperback 2009
ISBN: 978-0-86341-859-4 **£25**

Guidance Note 6: Protection Against Overcurrent, 5th Edition

Order book PWG6170B 104pp Paperback 2009
ISBN: 978-0-86341-860-0 **£25**

Guidance Note 7: Special Locations, 3rd Edition

Order book PWG7170B 144pp Paperback 2009
ISBN: 978-0-86341-861-7 **£25**

Guidance Note 8: Earthing & Bonding, 1st Edition

Order book PWRG0241 168pp Paperback 2007
ISBN: 978-0-86341-616-3 **£25**

continues overleaf ▶

Other guidance publications

Commentary on IEE Wiring Regulations (17th Edition, BS 7671:2008)
Order book PWR08640
c.432pp Hardback 2009
ISBN: 978-0-86341-966-9 **£65**

Electrical Maintenance, 2nd Edition
Order book PWR05100
228pp Paperback 2006
ISBN: 978-0-86341-563-0 **£40**

Code of Practice for In-service Inspection and Testing of Electrical Equipment, 3rd Edition
Order book PWR08630
152pp Paperback 2007
ISBN: 978-0-86341-833-4 **£40**

Electrical Craft Principles, Volume 1, 5th Edition
Order book PBNS0330
344pp Paperback 2009
ISBN: 978-0-86341-932-4 **£25**

Electrical Craft Principles, Volume 2, 5th Edition
Order book PBNS0340
432pp Paperback 2009
ISBN: 978-0-86341-933-1 **£25**

Electrician's Guide to the Building Regulations, 2nd Edition
Order book PWGP170B
234pp Paperback 2008
ISBN: 978-0-86341-862-4 **£22**

Electrical Installation Design Guide: Calculations for Electricians and Designers
Order book PWR05030
186pp Paperback 2008
ISBN: 978-0-86341-550-0 **£22**

Electrician's Guide to Emergency Lighting
Order book PWR05020
88pp Paperback 2009
ISBN: 978-0-86341-551-7 **£22**

Electrical training courses

We offer a comprehensive range of technical training at many levels, serving your training and career development requirements as and when they arise.

Courses range from Electrical Basics to Qualifying City & Guilds or EAL awards.

Train to the 17th Edition BS 7671:2008
▶ Update from 16th to 17th Edition
▶ Understand the changes
▶ New qualifying awards C&G/EAL
▶ Meet industry standards

Qualifying Courses
▶ Certificate of Competence Management of Electrical Equipment Maintenance (PAT) – 1 day
▶ Certificate of Competence for the Inspection and Testing of Electrical Equipment (PAT) – 1 day
▶ Certificate in the Requirements for Electrical Installations – 3 days
▶ Upgrade from 16th Edition achieved since 2001 – 1 day
▶ Certificate in Fundamental Inspection, Testing and Internal Verification – 3 days
▶ Certificate in Inspection, Testing and Certification of Electrical Installations – 3 days

Other 17th Edition Courses
▶ Earthing & Bonding – For designers and electrical contractors who require a good working knowledge of the E & B arrangements as required by BS 7671:2008
▶ 17th Edition Design – BS 7671 and the principles associated with the design of electrical installations

To view all our current courses and book online, visit
www.theiet.org/coursesbr

To discuss your training requirements and for on-site group training, please speak to one of our advisors on +44 (0)1438 767289

For more information, visit www.theiet.org/wiringregs

Notes

Notes